T0338626

SCIENCE IN HISTORY

Series Editors

Simon J. Schaffer, University of Cambridge

James A. Secord, University of Cambridge

Science in History is a major series of ambitious books on the history of the sciences from the mid-eighteenth century through the mid-twentieth century, highlighting work that interprets the sciences from perspectives drawn from across the discipline of history. The focus on the major epoch of global economic, industrial, and social transformations is intended to encourage the use of sophisticated historical models to make sense of the ways in which the sciences have developed and changed. The series encourages the exploration of a wide range of scientific traditions and the interrelations between them. It particularly welcomes work that takes seriously the material practices of the sciences and is broad in geographical scope.

Death in Beijing

In this innovative and engaging history of homicide investigation in Republican Beijing, Daniel Asen explores the transformation of ideas about death in China in the first half of the twentieth century. In this period, those who died violently or under suspicious circumstances constituted a particularly important population of the dead, subject to new claims by police, legal and medical professionals, and a newspaper industry intent on covering urban fatality in sensational detail. Asen examines the process through which imperial China's old tradition of forensic science came to serve the needs of a changing state and society under these dramatically new circumstances. This is a story of the unexpected outcomes and contingencies of modernity, presenting new perspectives on China's transition from empire to modern nation-state, competing visions of science and expertise, and the ways in which the meanings of death and dead bodies changed amid China's modern transformation.

Daniel Asen is Assistant Professor in the Department of History at Rutgers University, Newark. His published research has investigated the social and cultural contexts of science and medicine in late imperial and twentieth-century China, with attention to transnational and global perspectives. His work has been published in *Social History of Medicine* and *East Asian Science, Technology and Society*, as well as within edited volumes on legal history and the history of medicine in China. He is a member of the American Historical Association, the Association for Asian Studies, and the International Society for Chinese Law and History.

Death in Beijing

Murder and Forensic Science in Republican China

Daniel Asen

Rutgers University – Newark

CAMBRIDGE
UNIVERSITY PRESS

Shaftesbury Road, Cambridge CB2 8EA, United Kingdom

One Liberty Plaza, 20th Floor, New York, NY 10006, USA

477 Williamstown Road, Port Melbourne, VIC 3207, Australia

314–321, 3rd Floor, Plot 3, Splendor Forum, Jasola District Centre, New Delhi – 110025, India

103 Penang Road, #05–06/07, Visioncrest Commercial, Singapore 238467

Cambridge University Press is part of Cambridge University Press & Assessment, a department of the University of Cambridge.

We share the University's mission to contribute to society through the pursuit of education, learning and research at the highest international levels of excellence.

www.cambridge.org
Information on this title: www.cambridge.org/9781107126060

© Daniel Asen 2016

This publication is in copyright. Subject to statutory exception and to the provisions of relevant collective licensing agreements, no reproduction of any part may take place without the written permission of Cambridge University Press & Assessment.

First published 2016

A catalogue record for this publication is available from the British Library

Library of Congress Cataloging-in-Publication data
Asen, Daniel S., author.
Death in Beijing : murder and forensic science in Republican China / Daniel Asen (Rutgers University – Newark).
Cambridge, United Kingdom ; New York, NY : Cambridge University Press, 2016. | Series: Science in history
LCCN 2016017600| ISBN 9781107126060 (hardback) |
ISBN 9781107157160 (paperback)
LCSH: Murder – Investigation – China – Beijing – History – 20th century. | Forensic sciences – China – Beijing – History – 20th century. | Violent deaths – Social aspects – China – Beijing – History – 20th century. | Murder victims – China – Beijing – History – 20th century. | City and town life – China – Beijing – History – 20th century. | Beijing (China) – History – 20th century. | Death – Social aspects – China – History – 20th century. | Social change – China – History – 20th century. | China – History – Republic, 1912–1949. | BISAC: TECHNOLOGY & ENGINEERING / History.
LCC HV9960.C532 B34 2016 | DDC 363.25/95230951156–dc23
LC record available at https://lccn.loc.gov/2016017600

ISBN   978-1-107-12606-0   Hardback
ISBN   978-1-107-57160-0   Paperback

Cambridge University Press & Assessment has no responsibility for the persistence or accuracy of URLs for external or third-party internet websites referred to in this publication and does not guarantee that any content on such websites is, or will remain, accurate or appropriate.

Contents

Figures

Acknowledgments

It is a great pleasure to acknowledge everyone who made this book possible and to express my gratitude. While I wrote my dissertation and since completing it, Madeleine Zelin has provided an amazing level of encouragement and support for which I am truly grateful. Eugenia Lean also guided me through the process, providing essential feedback and support. Nancy Leys Stepan has encouraged this project in numerous ways, and it has benefited from our wide-ranging discussions. I am also thankful for intellectual exchanges with Dorothy Ko, who challenged my thinking about bodies, sources, and evidence. Lydia Liu, Zoë Crossland, and Rebecca Karl made this research richer through their interest, feedback, and encouragement. I am also grateful for Bill McAllister's support and the unique environment that he fostered in the Mellon Interdisciplinary Graduate Fellows Program. Li Chen has encouraged my work as well, and I am inspired by his efforts to promote the field of Chinese legal history. I also have to mention the friendship and support of Liza Lawrence, Arunabh Ghosh, Brian Tsui, and Tim Yang, as well as Andre Deckrow, Buyun Chen, Mi-Ryong Shim, and Michael McCarty. S.E. Kile and Greg Patterson deserve a special thanks for their support and comradery at the end. This research would not have been possible without the assistance of numerous individuals at Columbia's C.V. Starr East Asian Library, and I am especially thankful for the kind assistance of Ken Harlin.

My colleagues at Rutgers University-Newark and NJIT have been incredibly encouraging and supportive as I have brought this project to completion. For creating such a special academic and work environment, I have to thank Kornel Chang, Jon Cowans, Ruth Feldstein, Matthew Friedman, Eva Giloi, Jim Goodman, Mark Krasovic, Alison Lefkovitz, Lyra Monteiro, Maureen O'Rourke, Liz Petrick, Rabeya Rahman, Mary Rizzo, Beryl Satter, Richard Sher, Tim Stewart-Winter, Whit Strub, Nükhet Varlik, and Jennifer Arena. I also must express gratitude to the late Clem Price, whose deep kindness and generosity I experienced only briefly. I owe special thanks to Karen Caplan, Susan Carruthers, Gary

Farney, Stephen Pemberton, and Dean Jan Ellen Lewis for going above and beyond in the time and support that they have so generously given me, and also to Christina Strasburger, who has demonstrated amazing patience in showing me the ropes of the department and school. Being able to work on the book over 2014–15 while on research leave made all of the difference in completing the book manuscript, and I am thankful to have had this opportunity.

Many other individuals have encouraged my interest in this topic and supported the writing of my dissertation and this book. I am thankful for the support of colleagues whom I met at the University of Pittsburgh, especially Vincent Leung and John Stoner. I also thank Nathan Sivin and Paul R. Goldin for their encouragement at the University of Pennsylvania. This project owes much to David Luesink, whose many insights about dissection, professionalization, and modern Chinese history have been enlightening at every stage of the process. I thank David as well as Quinn Javers and Quentin (Trais) Pearson for their insightful feedback on my book manuscript. Early discussions with Thomas Buoye about Chinese law as well as ongoing discussions with Yi-Li Wu and Matthew Sommer about Chinese forensics have helped me to think through various issues in this project and my intellectual debts to them should be apparent in this book. I also thank Jeffrey Jentzen for his insights into the contemporary practice of forensic science, as well as its history, and for his deep generosity.

Over the course of three periods of research in Beijing I have benefited from the support of Dong Jianzhong, Guo Jinhai, Luo Zhitian, and Wang Lijun. I am also indebted to Li Nanchun, Liu Yi, Wang Xueshen, and Zhou Yuefeng for their generosity during my time in Beijing. My time at the Beijing Municipal Archives benefited immeasurably from the patience of Wu Heping and the rest of the archival staff, and I am grateful for their kindness and support. I am also thankful for the assistance of many individuals at the Capital Library, First Historical Archives of China, National Library of China, and library of Peking University and its Health Science Center. I also must thank the amazing staff of the Rockefeller Archive Center in Sleepy Hollow, NY.

This book also benefited from the feedback and comments of numerous individuals who participated in conferences, workshops, and talks at which I presented my work. Workshops and conferences such as "Medicine & Culture: Chinese-Western Medical Exchange (1644–ca. 1950)" at the Ricci Institute for Chinese-Western Cultural History (2007), "The (After)Life of Traditional Knowledge: The Cultural Politics and Historical Epistemology of East Asian Medicine" at University of Westminster (2010), the "Ritual as Technology in East

Asia Workshop" at the Max Planck Institute for the History of Science (2011), "Global Perspectives on Chinese Legal Medicine" at the University of Michigan (2011), the "International Workshop on Chinese Legal History/Culture/Modernity" at Columbia University (2012), "Social Lives of Dead Bodies in Late Imperial and Modern China" at Brown University (2013), and "Locating Forensic Science and Medicine" at University of Notre Dame Global Gateway, London (2015) were particularly important for the development of this project and I thank all of the organizers and participants.

I am also grateful for the encouragement of Lucy Rhymer at Cambridge University Press, whose support helped to move the book project forward and made it intellectually richer in the process, as well as the assistance of Rosalyn Scott and Claire Sissen. The comments of two anonymous reviewers were incredibly helpful and insightful, and the book benefited greatly from their feedback.

During the writing and research for my dissertation and book, I have received financial assistance from a number of sources, including a Henry Luce Foundation/ACLS Program in China Studies Postdoctoral Fellowship, an Andrew W. Mellon Foundation/ACLS Dissertation Completion Fellowship, funding and material support through the Paul F. Lazarsfeld Center at the Institute for Social and Economic Research and Policy (Columbia University), a Peking University Harvard-Yenching Fellowship for Advanced Research in Chinese Studies, summer and travel funding from the Weatherhead East Asian Institute (Columbia University), Department of History (Columbia University), and Department of East Asian Languages and Cultures (Columbia University), as well as a Fulbright US Student fellowship in 2004–2005 that made much of the initial research for this project possible. I am grateful for the generous support that I received from these organizations.

Portions of Chapters 2 and 3 appeared in the following publication, and I thank Brill for permission to include revised portions of this material in the book: Daniel Asen. "Old Forensics in Practice: Investigating Suspicious Deaths and Administering Justice in Republican Beijing." In *Chinese Law: Knowledge, Practice, and Transformation, 1530s to 1950s*, edited by Madeleine Zelin and Li Chen. Brill, 2015.

Finally, I am grateful for the steadfast support and encouragement of my family and Rachel throughout the process of researching and writing this book.

Introduction

On the afternoon of February 3, 1936, two metal boxes containing dismembered body parts were discovered at the East Station in Beiping, former capital of the Qing empire (1644–1911) and successive regimes of the early Republic and now a city that was falling under an ever-expanding Japanese military and political presence in north China. The boxes had been left at the station by a man who purchased a ticket on an evening train of the Beiping–Mukden railway line and fled the scene after leaving his cargo in the temporary care of station attendants. When the boxes were opened by police and station personnel, they were found to contain a dismembered body, divided into head and torso in one box and limbs in the other.[1] The city's Bureau of Public Safety (PSB) was subsequently notified, as was the city procuracy, an office responsible for investigating crime, overseeing prosecutions, and, as in this case, carrying out the forensic examination of bodies that had become the focus of homicide investigations.[2] Personnel from this office, including a procurator named Ming Yan and a forensic body examiner named Yu Depei, arrived at the scene and proceeded to examine the body parts in the presence of police and other officials. According to Yu, who inspected the body, the male victim had been wounded with a stick, hacked to death with a blade, and then dismembered after death.

An investigation led by city police began with interrogation of witnesses at the station and soon focused on what would turn out to be a mistaken identification of the body.[3] Less than two weeks later, the case had stalled and all clues had been exhausted.[4] It would be a month and a half before

---

[1] Beiping PSB to Outer Fifth District, February 11, 1936, Beijing Municipal Archives (BMA) J184-2-10790, 23–4; *Shibao* (SB) February 4, 1936, 4; *Shijie ribao* (SJRB) February 4, 1936, 6.

[2] The Beijing procuracy, or procurator's office, was an independent office known as the Capital Local Procuracy (Jingshi difang jianchating) until 1928, when it was reorganized as an office subordinate to the Beiping Local Court (Beiping difang fayuan).

[3] SB February 6, 1936, 4; SB February 7, 1936, 4.

[4] Beiping PSB to Outer Fifth District, February 18, 1936, BMA J184-2-10790, 30–2; SB February 23, 1936, 4.

the case was cracked, the murder solved by city detectives working under the supervision of senior detective Ma Yulin, whom the newspaper *Truth Post* characterized as an "Oriental Sherlock Holmes."[5] This happened in late March, and all details were subsequently revealed in the newspapers.[6] The investigation had turned up the case of a person from Henan named Zhang Shulin, whose disappearance had suspiciously not been reported to the authorities. Both Zhang and Wang Huayi, the head of the household in which Zhang had worked, had had sexual relations with a wet-nurse, also living in the household, named Mrs. Liu née Wang.[7] The resentment that Wang harbored from Zhang's continued contact with her had instigated the murder. Wang entrusted Li Liangjing, who performed various tasks for Wang, with killing Zhang Shulin with the assistance of another man in Wang's employ; the investigation also impli-cated several other members of the household who were eventually exon-erated. In the end, judges of the Beiping Local Court found Li guilty of murder, while Wang Huayi, who orchestrated the killing and dismember-ment, was never apprehended.[8]

As in the case of other unidentified bodies found in the city, police authorities had provided a coffin and buried the body in one of the city's public burial grounds soon after it was discovered.[9] Yet, those who followed the intensive coverage that this case received in the newspapers might have noted, months later, a seeming aberration in the burial of Zhang's then-unidentified body: the police had never buried the head.[10] In fact, almost immediately after the discovery of Zhang's body, the city PSB had sent the head to the department of legal medicine of the medical school of Beiping University, where it was preserved and examined on at least two occasions between February and late April 1936, when officials of the procuracy requested an additional examination to confirm whether the head had been hacked off while the victim was alive or after death.[11]

---

[5] SB March 23, 1936, 4. For more on the translation of Sherlock Holmes in early twentieth-century China and other appearances of the appellation "the Oriental Holmes," see Jeffrey Kinkley, *Chinese Justice, the Fiction: Law and Literature in Modern China* (Stanford: Stanford University Press, 2000), 399 n45.

[6] SB March 23, 1936, 4; SJRB March 23, 1936, 6.

[7] Official documents and news reports describing the course of legal cases such as this one generally identified women by their father's surname or, if married, by the surnames of both husband and father, followed by the character *shi*. As in this case, I refer to married women who are identified in this way as Mrs. Liu née Wang (originally "Liu Wang shi").

[8] For the outcome of the trial in the Beiping Local Court, see the court's April 1937 judgment, published in SJRB May 12–17, 1937, 5.

[9] SJRB February 5, 1936, 6.

[10] This was clear from the victim's father's petition to have the body reburied three months later, printed in SJRB May 9, 1936, 6. Also see SB May 9, 1936, 4.

[11] "Beiping dongchezhan xiangshi an zhi jianding" [The appraisal in the Beiping East Station box-corpse case], *Beiping yikan* 4, no. 7 (1936): 59–63; SJRB May 14, 1937, 5.

These examinations were carried out by Lin Ji (G. Lynn, 1897–1951), the director of the department and an expert in legal medicine who over the previous decade had played an important role in introducing this new field of scientific expertise to Chinese courts and establishing institutional foundations for the discipline in China.

In the autopsy room of this facility, Lin Ji inspected the skin, muscles, and blood vessels at the place where Zhang Shulin's head had been severed from the rest of his body. This examination yielded a lack of hemorrhaging or other "vital reaction" (*shenghuo fanying*) that would indicate that Zhang had still been alive when the cuts were made, thus confirming for police and the procuracy that the decapitation had taken place after death. Lin Ji's inspection of the head and neck also suggested that death had been caused by strangulation and not from the wounds inflicted by a blade; ligature marks were discovered at the nape of the neck and around its sides, and these appeared to have been inflicted before death. That these findings were guaranteed by modern Western science (*kexue*), a form of knowledge that had attained tremendous intellectual authority in China over the previous decades, was made explicit in the phrase that Lin Ji used to conclude this and every other examination report that he sent to legal officials: "The above explanation is made according to scientific theory and the facts" (*you shuoming xi ju xueli shishi*).[12] That is, the findings accorded with the formal knowledge of the sciences and the powerful epistemological authority with which they were associated.

The examination report that Lin Ji returned to the procuracy with his findings was a detailed, thorough document, one that made a strong statement about the power of medico-legal experts such as Lin Ji to bring new fields of scientific knowledge to bear on physical evidence. Yet, if it is read against the grain one can detect as well ominous traces of the limitations that had been placed on the explanatory power of this new form of forensic expertise as it was practiced in Beiping during this period. Whether by intention or simply neglect, police and legal officials had declined to send the rest of Zhang's body for examination, a decision that, Lin Ji noted in the report, made it impossible for him to conclusively determine the cause of death – one of the most important questions in a case such as this one.[13] That this oversight had occurred in the first place

---

[12] For more on the meaning and significance of *xueli* (theory, scientific principle) as an epistemological concept in contemporary discourses of forensic knowledge, see Daniel Asen, "The only options?: 'Experience' and 'theory' in debates over forensic knowledge and expertise in early twentieth-century China," in *Historical Epistemology and the Making of Modern Chinese Medicine*, ed. Howard Chiang (Manchester: Manchester University Press, 2015), 143–5.

[13] "Beiping dongchezhan xiangshi an zhi jianding," 61.

was a result of the fact that the process of homicide investigation in Beiping was not primarily organized around the involvement of the medico-legal laboratory, nor did medical experts such as Lin Ji have a strong legal or physical claim over the dead body itself. As in this case, the more consequential forensic examinations usually occurred under the authority of the city procuracy at the start of the police investigation. Moreover, these examinations, performed by inspection clerks (*jianyan li*) of the procuracy such as Yu Depei, were carried out in a way that implicitly challenged the authority of China's nascent legal medicine profession: the body examined in these instances was not a body of muscles and tissues – that is, a conception of the body informed by Western anatomy – but rather one that relied on an alternative system of bodily knowledge, based not on the globally authoritative knowledge of the modern sciences, but on forensic practices that had developed under the legal system of the recently fallen Qing empire.

The discovery of Zhang Shulin's body, much like other cases involving suspicious death or murder, presented problems of knowledge: What was the cause of death? If homicide, who should be held legally responsible? Addressing these questions necessarily meant raising other questions about who was competent to interpret the meaning of physical evidence and the sources of authority on which such decisions about bodies and things, so consequential for the pursuit of justice, should be made. Investigating homicide also raised questions about how to reconcile the imperative, rooted in both law and science, to make the dead body accessible to officials and experts with the other social meanings and ritual imperatives that already shaped the cultural terrain of death in early twentieth-century Beijing. This book is about the ways in which officials, professionals, and city people negotiated these questions against the immediate backdrop of a city undergoing modern transformation and within the larger context of the collapse of the Qing, a sophisticated early modern empire that left complex legacies in its wake.

Homicide Investigation and the Beijing Procuracy

There were many occupations in early twentieth-century Beijing whose members played the role of "death brokers," in the words of sociologist Stefan Timmermans, as "expert intermediaries who establish the varying meanings of violent or suspicious death."[14] Police detectives, hygiene officials, ritual specialists, journalists, medical school pathologists, and

---

[14] Stefan Timmermans, *Postmortem: How Medical Examiners Explain Suspicious Deaths* (Chicago: The University of Chicago Press, 2006), 251.

forensic scientists such as Lin Ji all played roles in explaining the causes and implications of death for family, society, and the state. For the particular population of those who died violently or under suspicious circumstances, few of these were as important as the officials of the Beijing procuracy, an office that was established as a cornerstone of the new legal system that was adopted in China during the New Policies reforms (1901–11), a significant period of institutional reform and innovation that had transformed the legal and political structures of the Qing. Much as in imperial Russia's legal reforms of the 1860s or those that followed the Meiji Restoration in Japan (1870s–1890s), reformers of the Qing empire drew heavily on continental European models of judicial procedure when formulating this new legal system.[15] Like other aspects of the judicial reform that was carried out during these years, the conception of the office of the procuracy, its professional authority, and relations with police and other actors in the law were all informed by Japanese models, themselves the product of two decades of experimentation with French and German legal institutions.[16]

In early twentieth-century China, as in other modern countries that have adopted the procurator system, this office and its staff played a very direct role in establishing the facts of criminal cases prior to trial, often by interrogating suspects and witnesses and performing other tasks that many legal systems delegate to police detectives.[17] The important role that Chinese procurators played in the investigation of homicide cases was thus an area of their work with obvious global precedents. At the same time, procurators in China were given unique authority over the forensic examination of the dead body, a task referred to as *xiangyan* or *jianyan* in Chinese. While both words have the general meaning of "inspection," these were words that had a long history in imperial China as technical terms for the forensic inspection of a body.[18] In the case of the

---

[15] Elisa M. Becker, *Medicine, Law, and the State in Imperial Russia* (Budapest: Central European University Press, 2011); Wilhelm Röhl, *History of Law in Japan since 1868* (Leiden: Brill, 2005).

[16] Röhl, *History of Law in Japan Since 1868*, 789–93; Xie Rucheng, *Qingmo jiancha zhidu ji qi shijian* [The late Qing procuratorial system and its practice] (Shanghai: Shanghai renmin chubanshe, 2008), 23–31, 474–84. For the important role that Japanese experts, institutional models, and knowledge played in these reforms more generally, see Douglas R. Reynolds, *China, 1898–1912: The Xinzheng Revolution and Japan* (Cambridge: Council on East Asian Studies, Harvard University. Harvard University Press, 1993).

[17] Becker, *Medicine, Law, and the State in Imperial Russia*, 195–8; David T. Johnson, *The Japanese Way of Justice: Prosecuting Crime in Japan* (Oxford: Oxford University Press, 2002), 51–4.

[18] The meanings of both terms were analyzed and debated in a number of late imperial works on local administration. Xin-zhe Xie, "Procedural Aspects of Forensics Viewed through Bureaucratic Literature in Late Imperial China" (Paper prepared for "Global Perspectives on the History of Chinese Legal Medicine," University of Michigan, Ann

dismembered body discovered at the East Station, as in other cases of suspicious death or homicide, the forensic examination of the body was carried out early in the investigation by this office, often preceding the involvement of detectives in the case. These examinations were not, strictly speaking, supposed to be carried out by physicians: the laws and institutional procedures that defined the relationship between law and medicine as professions gave officials of the judiciary tremendous authority over the forensic examination of bodies. Indeed, it was not unusual for procurators to oversee the initial examination of a body without any assistance from physicians and to decide for themselves if, when, and how to involve medical experts such as Lin Ji in the case.

The technical knowledge that Chinese procurators and their staff used to examine the corpse and find cause of death in such cases had a long history. While there are traces of evidence for sophisticated methods of criminal investigation and forensic examination throughout the early imperial period of Chinese history and even in the legal procedure of the Qin state prior to its unification of the empire in 221 B.C., an important turning point in the development of forensic knowledge in China was the completion of the *Collected Writings on the Washing Away of Wrongs* (Xiyuan jilu), a detailed handbook of forensic examination authored by an experienced local administrator and judicial official of the Southern Song (1127–1279) named Song Ci (1186–1249).[19] This text and its subsequent late imperial editions contained practical instructions which officials could use when investigating cases of suspected homicide and non-fatal injury, including detailed descriptions of the bodily signs that could be expected in fatal beating, strangulation, poisoning, and myriad other scenarios. Under the Qing empire, a highly bureaucratized system of forensic examination was established on the basis of later versions of the *Washing Away of Wrongs* as well as detailed regulations governing all aspects of the administration of forensic examinations.[20] This arrangement was not based, foremost, on the

Arbor, October 2011), 15–24. Generally, *xiangyan* was used to refer to the initial examination of a "fresh" corpse – not skeletal remains – while *jianyan* could refer to both. Both usages were common during the Republican period.

[19] Jia Jingtao, *Zhongguo gudai fayixue shi* [A history of legal medicine in ancient China] (Beijing: Qunzhong chubanshe, 1984); Nathan Sivin, ed., *Science and Civilisation in China. Volume 6: Biology and Biological Technology. Part VI: Medicine*. By Joseph Needham with the collaboration of Lu Gwei-djen (Cambridge: Cambridge University Press, 2000), Part E, "Forensic medicine," 175–200; Derk Bodde, "Forensic Medicine in Pre-Imperial China," *Journal of the American Oriental Society* 102, no. 1 (1982): 1–15.

[20] Jia, *Zhongguo gudai fayixue shi*, 105–21; Pierre-Étienne Will, "Developing Forensic Knowledge through Cases in the Qing Dynasty," in *Thinking with Cases: Specialist Knowledge in Chinese Cultural History*, ed. Charlotte Furth, Judith T. Zeitlin, and Ping-chen Hsiung (Honolulu: University of Hawai'i Press, 2007); Xin-zhe Xie,

institutionally recognized role of outside experts – physicians or otherwise – in the law, a significant contrast with legal procedure in early
modern continental Europe, which heavily supported physicians' involvement in legal cases while fostering the development of a significant body
of forensic knowledge.[21]

The overarching goal of forensic procedure in late imperial China was,
rather, to standardize the techniques that local officials used to collect,
document, and analyze the body evidence, a reflection of the Qing judicial
system's high degree of procedural routinization in the investigation and
adjudication of homicide cases more generally.[22] In these ways, the
development of forensic knowledge in late imperial China was inseparable from a broader institutional impetus to regulate the actions of the
bureaucracy's officials and minor functionaries in this important stage of
legal proceedings. For this reason, higher-level officials in the provinces
and central government were supposed to possess enough forensic knowledge to evaluate the integrity of the forensic examinations conducted in
local cases; likewise, county officials were supposed to possess enough
forensic knowledge to supervise the *wuzuo*, a term with unclear meaning
and unknown provenance (at times written with the alternate character
wu) that was used to refer to the local functionaries who actually examined the body in such cases.[23]

In English-language scholarship this term is usually rendered as "coroner" or "ostensor," the latter indicating the role that the *wuzuo* played in
pointing out the wounds and other forensic signs for supervising
officials.[24] The forensic role that *wuzuo* played as subordinate judicial

"The Shaping of Autopsy Evidence in Nineteenth-Century China" (Paper prepared for
"The Social Lives of Dead Bodies in Modern China," Brown University, June 2013);
Chen Chong-Fang, "'Xiyuan lu' zai Qingdai de liuchuan, yuedu yu yingyong"
[The Circulation, Reading, and Using of Xiyuan-lu in Qing Dynasty], *Fazhi shi yanjiu*
25 (2014): 37–94.

[21] Catherine Crawford, "Legalizing Medicine: Early Modern Legal Systems and the
Growth of Medico-legal Knowledge," in *Legal Medicine in History*, ed. Michael Clark
and Catherine Crawford (Cambridge: Cambridge University Press, 1994); Silvia De
Renzi, "Witnesses of the Body: Medico-legal Cases in Seventeenth-century Rome,"
Studies in History and Philosophy of Science 33 (2002): 219–42.

[22] Thomas Buoye, "Suddenly Murderous Intent Arose: Bureaucratization and
Benevolence in Eighteenth-century Qing Homicide Reports," *Late Imperial China* 16,
no. 2 (1995): 62–97.

[23] For brief discussions of the term itself, see Chang Che-chia, "'Zhongguo chuantong
fayixue' de zhishi xingge yu caozuo mailuo" [Knowledge and Practice in "Traditional
Chinese Forensic Medicine"], *Zhongyang yanjiuyuan jindaishi yanjiusuo jikan* 44 (2004):
11–2; Jia, *Zhongguo gudai fayixue shi*, 59; Yang Fengkun, "'Wuzuo' xiaokao" [A brief
inquiry into the "wuzuo"], *Faxue* 7 (1984): 40–1. Also see Luo Zhufeng, *Hanyu dacidian*
(Shanghai: Shanghai cishu chubanshe, 2008), v. 1, 1195 and v. 7, 432.

[24] For the latter, see Sivin, ed., *Science and Civilisation in China*, 191. I have used the word
"coroner" in my previous published work: i.e., Asen, "The only options?."

assistants differed significantly from that of Anglo-American coroners, who were elected or appointed officials who oversaw investigations into violent or otherwise suspicious deaths on their own authority.[25] Given that these were, roughly speaking, the very functions performed by the late imperial county magistrates and early twentieth-century procurators under whom the *wuzuo* worked, Brian McKnight's use of "coroner's assistant" as a translation for this position provides perhaps a more accurate description of the role that they actually played.[26] We might also compare the function of the *wuzuo* with that of the judicial investigators who assisted late nineteenth-century Russian procurators in their criminal investigations.[27] This is a particularly fitting historical comparison given that the position of "inspection clerk" (*jianyan li*), which was created under the New Policies reforms to replace the old position of *wuzuo*, was established from the outset as a subordinate functionary of procuratorial officials.[28] Inspection clerks in Beijing worked closely, we will see, with judicial police, who also worked under the direction of the procuracy.

This arrangement of subordinate body examiners and supervising officials, and the particular way of distributing technical knowledge on which it was based, was quite successful. By the end of the dynasty, case files reveal a high level of consistency in forensic examination procedures and in the terminology used to describe bodies and wounds.[29] In the case of the dismembered body discovered at the East Station (and many other cases as well), it was this system of institutionalized knowledge that Republican judicial officials relied upon to solve and adjudicate some of the most serious crimes in the law codes. During a period in which the modern social sciences, Western scientific medicine and public health, and other fields of scientific and professional expertise were adopted by the modernizing Chinese state, it is striking that this old field of technical

---

[25] Ian A. Burney, *Bodies of Evidence: Medicine and the Politics of the English Inquest, 1830–1926* (Baltimore: The Johns Hopkins University Press, 2000); Jeffrey M. Jentzen, *Death Investigation in America: Coroners, Medical Examiners, and the Pursuit of Medical Certainty* (Cambridge: Harvard University Press, 2009), 9–30.

[26] Brian E. McKnight, *The Washing Away of Wrongs: Forensic Medicine in Thirteenth-Century China* (Ann Arbor: Center for Chinese Studies, The University of Michigan, 1981), 11.

[27] Becker, *Medicine, Law, and the State in Imperial Russia*, 196–8.

[28] There was a great deal of slippage between the terms *wuzuo* and "inspection clerk" during the Republican period. The former never disappeared from official or popular lexicons and the latter remained indeterminate given that no uniform standard governing who could serve as an inspection clerk was ever enforced. Likewise, some old *wuzuo* – that is, individuals who had served in this position under the Qing – became inspection clerks.

[29] For some of the basic procedures used to examine and document wounds, see Daniel Asen, "Vital Spots, Mortal Wounds, and Forensic Practice: Finding Cause of Death in Nineteenth-Century China," *East Asian Science, Technology and Society: An International Journal* 3, vol. 4 (2009): 453–74.

knowledge continued to receive the endorsement of Chinese officials in local, provincial, and central government offices for decades after the fall of the Qing.[30] Despite the regional fragmentation and other challenges of modern state-building and judicial reform during the Republican period, the officials who staffed China's modern courts and local judicial offices used these old techniques to maintain a surprising degree of uniformity across homicide investigation practices in widely divergent geographic locations, ranging from urbanized Shanghai and Beijing to counties that lacked modern courts. Moreover, these techniques were used with widely acknowledged effectiveness to examine the body and find cause of death in all different kinds of cases, whether the routine discovery of dead bodies on the streets, sensational murders such as the case of Zhang Shulin, or cases that had even greater implications, such as the massacre of protestors by Duan Qirui's (1865–1936) government on March 18, 1926, one example of the escalating violence and mass politics that were reshaping the political landscape of Beijing during this period.[31]

Legal Medicine and the Rise of the Modern Professions in China

In his study of changes in the organization of handicraft production and skill among Sichuan papermakers, Jacob Eyferth has argued that

the Chinese revolution – understood as a series of interconnected political, social, and technological transformations – was as much about the redistribution of skill, knowledge, and technical control as it was about the redistribution of land and political power, and that struggles over skill in twentieth-century China resulted in a massive transfer of technical control from the villages to the cities, from primary producers to managerial elites, and from women to men.[32]

Eyferth is primarily concerned with the process through which a modern technocratic state and new groups of elites formalized and appropriated handicraft skills under the imperatives of industrial development. Yet, his point very much applies to other twentieth-century shifts in the distribution and use of technical knowledge and the organization of work that

---

[30] Ruth Rogaski, *Hygienic Modernity: Meanings of Health and Disease in Treaty-Port China* (Berkeley: University of California Press, 2004); Tong Lam, *A Passion for Facts: Social Surveys and the Construction of the Chinese Nation-state, 1900–1949* (Berkeley: University of California Press, 2011).

[31] For discussion of the massacre within these larger contexts, see David Strand, *Rickshaw Beijing: City People and Politics in the 1920s* (Berkeley: University of California Press, 1989), 182–97.

[32] Jacob Eyferth, *Eating Rice from Bamboo Roots: The Social History of a Community of Handicraft Papermakers in Rural Sichuan, 1920–2000* (Cambridge: Harvard University Asia Center. Distributed by Harvard University Press, 2009), 1–2.

occurred under the influence of new professions and changing conceptions of occupational expertise. In law, medicine, and other fields, the rise of the modern professions signaled changing expectations about who in Chinese society was competent to handle particular occupational tasks, what counted as the most authoritative forms of expertise, and how an increasingly complex division of labor should be organized.

The rise of the modern professions in China was foremost a process of social change in which numerous areas of occupational and governmental work, including the investigation of homicide, came under the new influence of individuals who were generally distinguished from the members of other occupations (and within society more generally) by experience with higher education, higher incomes, and corresponding patterns of consumption and leisure, a reflection of the relatively higher socioeconomic status that they enjoyed in the changing urban economy.[33] Unsurprisingly, such individuals tended to have experience with Euro-American or Japanese institutions and knowledge acquired through study abroad, advanced post-graduate training in other countries, or simply from reading foreign literature in their fields of expertise. Most importantly, such individuals tended to utilize the knowledge associated with internationally recognized academic disciplines to claim authority in their occupational work, at times by establishing institutionalized restrictions such as licensing requirements that were meant to limit membership to those who had received specialized training and guarantee professionals' access to particular areas of work.[34] An important element of these new kinds of authority-claims was an association between professionals'

---

[33] Xiaoqun Xu, *Chinese Professionals and the Republican State: The Rise of Professional Associations in Shanghai, 1912–1937* (Cambridge: Cambridge University Press, 2001), 23–77. In this book, these individuals are referred to as members of modern "professions" or simply as "professionals," terms used here primarily as analytical categories, not actors' categories. While, as Xiaoqun Xu notes, *ziyou zhiye zhe* (free professionals) was an official classification endorsed by the Nationalist state in 1929, the term was not commonly used in materials pertaining to forensics. For general discussion of the sources of modern professionals' authority, see Eliot Freidson, *Professional Powers: A Study of the Institutionalization of Formal Knowledge* (Chicago: The University of Chicago Press, 1986); Andrew Abbott, *The System of Professions: An Essay on the Division of Expert Labor* (Chicago: The University of Chicago Press, 1988).

[34] Chinese professionals' attempts to create such restrictions saw uneven successes, a story that can be followed in Xu, *Chinese Professionals and the Republican State*. For example, during this period there were no effective, institutionally enforced barriers to becoming a physician of Western scientific medicine or of Chinese medicine. Lawyers, by contrast, were relatively more effective at controlling access to their profession. Bridie Andrews observes this point as well, following discussion of the limited successes of medical licensing efforts. Bridie Andrews, *The Making of Modern Chinese Medicine, 1850–1960* (Vancouver: UBC Press, 2014), 150–6. For the ways in which struggles over legal credentials played out in Beijing, see Michael Ng's study of the Beijing Bar Association's attempts to ban the practice of so-called phony lawyers (*fei lüshi*): Michael

expertise and modern Western science, a form of knowledge that was recognized to be of broad relevance across a range of intellectual and political agendas and thus gained a great amount of authority during this period.[35]

These shifting patterns of occupational expertise and social power affected Chinese practices of homicide investigation in different ways. Most immediately, Chinese physicians of Western scientific medicine began to assert a direct role in the forensic examination of bodies and other physical evidence. This was one area among many in which medical professionals began to make new kinds of claims over Chinese bodies for purposes of healthcare, academic research in anatomy and other fields, and the new broad-ranging imperative to investigate the health and vitality of the "population" (renkou), now conceived as a biological entity.[36] The influence of Chinese physicians in homicide investigation began in limited ways during the 1910s, expanded in the 1920s, and gained significant momentum during the Nanjing decade (1927–37) with the Nationalist party-state's active support for legal medicine, a discipline that had originated in early modern continental Europe and was established in the universities and state organs of a number of countries during the late nineteenth and early twentieth centuries.[37]

There were, to be sure, precedents from the late imperial period for Chinese physicians to play a role in forensic examinations and for the multi-directional exchange of specialist knowledge between medicine and forensics. As Yi-Li Wu has shown, for example, scholar-physicians in nineteenth-century China drew on forensic knowledge of the vital points

H.K. Ng, *Legal Transplantation in Early Twentieth-Century China: Practicing Law in Republican Beijing (1910s-1930s)* (London: Routledge, 2014), 78–84.

[35] D.W.Y. Kwok, *Scientism in Chinese Thought, 1900–1950* (New Haven: Yale University Press, 1965).This reflects a general pattern in the strategies that modern professionals use to legitimize their expertise – namely, by invoking authoritative modern values such as "rationality, logic, and science" as a "symbolic" resource for claiming authority. Abbott, *The System of Professions*, 54.

[36] David Luesink, "Dissecting Modernity: Anatomy and Power in the Language of Science in China" (PhD dissertation, University of British Columbia, 2012); Malcolm Thompson, "Foucault, Fields of Governability, and the Population—Family–Economy Nexus in China," *History and Theory* 51, no. 1 (2012): 42–62.

[37] When the English term "legal medicine" is used in this book, it is usually used in reference to the Chinese word *fayixue*, which was a new disciplinary term that was introduced into Chinese from the Japanese word *hōigaku*. In early twentieth-century China, much as in Japan during this period, "legal medicine" was much more commonly used than "forensic medicine" (*caipan yixue*) or "medical jurisprudence," both of which are other names for academic disciplines that are concerned, broadly speaking, with the inter-professional intersection of medicine and law, usually in order to place the expert knowledge of the former in service of the latter. "Forensic science," which was not a term that saw use in early twentieth-century China, is used in this book to refer generally to scientific disciplines, including legal medicine, that are oriented toward legal problems.

and used bodily knowledge contained in the *Washing Away of Wrongs* more generally in their discussions of trauma medicine and other fields of medical knowledge.[38] Yet, the new calls for establishing professional medical authority over forensic investigation that emerged during the early twentieth century were qualitatively different from earlier interactions between medicine and law in China. They must be located, specifically, in relation to a global shift through which a new set of forensic practices, institutions, and bodies of knowledge that had developed in continental Europe and Britain gained authority in many other societies, both colonies and independent states, throughout Asia, Africa, and the Americas.[39]

The forces that drove this process of global change in forensic practices were complex. In nineteenth-century America, Britain, and Russia, physicians asserted a greater role for medicine in the law in order to strengthen the public authority of medical experts more generally while contributing to broader moments of political and legal reform.[40] For some societies, modern forensic medicine was introduced under a new set of unequal power relations defined by Western imperialism and the challenges to political autonomy and sovereignty presented by these new circumstances. In early twentieth-century China, much as in nineteenth-century Khedival Egypt and Siam, political leaders and elites supported the introduction of new forensic practices as part of a broader package of policing, judicial, and medical institutions that were meant to strengthen the state in part as a response to these new pressures and challenges.[41]

---

[38] Yi-Li Wu, "Between the Living and the Dead: Trauma Medicine and Forensic Medicine in the Mid-Qing," *Frontiers of History in China* 10, no. 1 (2015): 38–73; Yi-Li Wu, "Bodily Knowledge and Western Learning in Late Imperial China: The Case of Wang Shixiong (1808–68)," in *Historical Epistemology and the Making of Modern Chinese Medicine*, ed. Howard Chiang (Manchester: Manchester University Press, 2015), 88–9, 95.

[39] This is not to diminish the importance of colonial sites for the development of forensic practices or the importance of colony-to-metropole (or colony-to-colony) transfers of techniques and knowledge, a pattern that is crucially important in the history of fingerprinting. This story can be followed in Simon A. Cole, *Suspect Identities: A History of Fingerprinting and Criminal Identification* (Cambridge: Harvard University Press, 2002), 60–96. For broad perspectives on the global history of forensic science during the modern period, see Jeffrey M. Jentzen, "Death and Empire: Legal Medicine in the Colonization of India and Africa," in *Medicine and Colonialism: Historical Perspectives in India and South Africa*, ed. Poonam Bala (London: Pickering & Chatto, 2014); Jia Jingtao, *Shijie fayixue yu fakexue shi* [The world history of legal medicine and sciences] (Beijing: Kexue chubanshe, 2000).

[40] James C. Mohr, *Doctors and the Law: Medical Jurisprudence in Nineteenth-Century America* (Baltimore: The Johns Hopkins University Press, 1993); Burney, *Bodies of Evidence*; Becker, *Medicine, Law, and the State in Imperial Russia*.

[41] Khaled Fahmy, "The Anatomy of Justice: Forensic Medicine and Criminal Law in Nineteenth-Century Egypt," *Islamic Law and Society* 6, no. 2, The Legal History of Ottoman Egypt (1999): 224–71; Quentin A. Pearson, "Bodies Politic: Civil Law &

The rapid changes in Chinese society that gave rise to a legal medicine profession during the first decades of the twentieth century were thus inseparable from social, political, economic, and intellectual changes occurring at a global level that invested European-derived scientific disciplines with authority and legitimacy in many societies during this period.[42]

Legal medicine was introduced to China during a period in which forensic knowledge and the organization of forensic practice were undergoing rapid and consequential shifts all across the globe. One can trace these shifts, at one level, through a succession of new technical procedures and scientific discoveries that significantly expanded the capabilities of expert-based forensic investigation and made a broader range of physical evidence intelligible to the forensic gaze. These developments included new tests for confirming poisoning by arsenic and other toxins, new tests for confirming the presence of blood and distinguishing between human and animal blood, a revolution in procedures of individual identification associated with Bertillonage and fingerprinting, and the emergence of the crime scene as a source of forensic evidence which, if preserved and documented in disciplined ways, could yield unprecedented insight into the course of a crime.[43] Not surprisingly, these changes in the practice of forensic investigation coincided with a growing disciplinary specialization within the forensic sciences and the development of a more complex division of labor within homicide investigation – shifts that were themselves inseparable from the expanding role of the laboratory in different forensic disciplines and the increasingly inaccessible nature of forensic testing protocols and knowledge-claims to non-specialists.[44] As a result, an important subtext of late nineteenth- and early twentieth-century forensic controversies in Britain, China, and elsewhere was the extent to which laypersons should defer to forensic

Forensic Medicine in Colonial Era Bangkok" (PhD dissertation, Cornell University, 2014).

[42] Cyrus Schayegh, *Who Is Knowledgeable Is Strong: Science, Class, and the Formation of Modern Iranian Society, 1900–1950* (Berkeley: University of California Press, 2009).

[43] For a general overview of developments in forensic science in the West during the modern period, see Katherine D. Watson, *Forensic Medicine in Western Society: A History* (New York: Routledge, 2011). For the history of new identification procedures, see Cole, *Suspect Identities*. For the history of crime scene investigation as a new and distinctive paradigm of forensic inquiry, see Ian Burney and Neil Pemberton, *Murder and the Making of English CSI* (Baltimore: The Johns Hopkins University Press, 2016).

[44] For the importance of specialization in the twentieth-century forensic sciences, see Watson, *Forensic Medicine in Western Society*, 125, 131–4. For historical perspectives on the intellectual, social, and institutional contexts of modern laboratory science and its rising prominence in nineteenth- and twentieth-century medical disciplines in the West, see Andrew Cunningham and Perry Williams ed., *The Laboratory Revolution in Medicine* (Cambridge: Cambridge University Press, 1992).

specialists – whether generalist physicians, pathologists, or experts in toxicology – when interpreting the physical evidence in a criminal inquiry or trial.[45]

Despite the powerful new technical capabilities of the forensic sciences during this period, in early twentieth-century China there was a persistent tension between medical experts' claims of indispensability in the law and the inconsistent demands for scientific expertise that were actually generated under the particular institutional practices and needs of the Republican judiciary.[46] The difficulties that China's first generations of professional forensic scientists faced in asserting authority in homicide investigation were not simply the result of legal officials' conservative refusal to adopt science-based forensic practices, a claim often made by proponents of forensic reform. Rather, a model of forensic expertise based on the establishment of urban-based laboratories and the deployment of highly trained medical specialists was, in many ways, ill-suited to the actual needs and resources of the Chinese judiciary of the 1920s and 1930s, a sprawling bureaucracy in which officials routinely administered justice in localities that lacked the financial resources to augment their forensic capabilities with additional specialist personnel or laboratory facilities.[47] Moreover, judicial officials' employment of subordinate body examiners allowed them to maintain direct authority over the collection and interpretation of forensic evidence, a practical arrangement that facilitated the investigation process in many of the cases handled by the Beijing procuracy and other judicial offices. The fact that homicide cases were often successfully solved without the use of laboratory testing or even a physician's autopsy also meant that physicians' claims of

---

[45] Burney, *Bodies of Evidence*, 107–36; Ian Burney, *Poison, Detection, and the Victorian Imagination* (Manchester: Manchester University Press, 2006), 49–53, 130–44.

[46] For more on the important relationship between the market for professionals' expertise and patterns of training and employment, see Elisabeth Köll, "The Making of the Civil Engineer in China: Knowledge Transfer, Institution Building, and the Rise of a Profession," in *Knowledge Acts in Modern China: Ideas, Institutions, and Identities*, ed. Robert Culp, Eddy U, and Wen-hsin Yeh (Berkeley: Institute for East Asian Studies Publications, forthcoming), which examines the impact of early twentieth-century China's semi-colonial pattern of railroad development on the professional structures and career prospects of engineers.

[47] For the challenges of financing local judicial institutions, see Xiaoqun Xu, *Trial of Modernity: Judicial Reform in Early Twentieth-Century China, 1901–1937* (Stanford: Stanford University Press, 2008), 149–83. The problem of how to distribute Western medicine-based technologies across a population that was largely agrarian and overwhelmingly poor was one that preoccupied promoters of rural healthcare such as PUMC graduate Chen Zhiqian (C.C. Chen) and his colleagues, who devised measures for the employment of rural health workers who could handle first aid, vaccinations, and other basic medical services. Mary Brown Bullock, *An American Transplant: The Rockefeller Foundation and Peking Union Medical College* (Berkeley: University of California Press, 1980), 162–72, 182–9.

forensic indispensability in homicide investigation were tempered by the redundancy of their expertise in practice. Within this context, the old forensic techniques of the late imperial state continued to be used in Beijing and elsewhere in China for many decades after the fall of the Qing despite the growing criticisms of proponents of Western legal medicine.

Ultimately, as legal and medical professionals negotiated their competing claims over homicide investigation, the authority that each attained over this work was far from absolute.[48] Chinese judicial officials maintained a strong professional jurisdiction over the forensic inspection of corpses, often to the disadvantage of legal medicine experts such as Lin Ji. At the same time, there were many instances over the 1920s and 1930s when physicians served as the occasional outside expert in legal proceedings or, in some cases, played an even greater role in the everyday operations of urban judicial offices. These negotiations over the practical work of homicide investigation took place within the context of much broader cultural and intellectual shifts that invested scientific knowledge with tremendous authority, a process that began during the last decade of Qing rule. In the process, new norms of medical knowledge, expertise, and proof did become important to the judiciary's own forensic work, often in ways that preceded the actual involvement of physicians in legal cases. Thus, officials might retrain inspection clerks in basic Western science as a way of improving their forensic knowledge and skills; likewise, the *Washing Away of Wrongs*, or particular examination practices associated with it, might be revised to better accord with the norms of legal medicine itself, a direction for forensic reform that was broadly accepted within the judiciary by the mid-1930s.[49] The end result of these negotiations was that the judiciary-centered model of forensics that the Republican state inherited from the Qing remained in place even as many in China, including judicial officials themselves, increasingly accepted the epistemological norms of the new science-based forensics and the basic assumption that professional medical experts should play a role in homicide investigation.

---

[48] For general discussion of the ways in which modern professional groups attempt to reorganize divisions of labor within society and the practical "settlements" that result from these negotiations, see Abbott, *The System of Professions*, 69–77 especially.

[49] Compare with Tina Phillips Johnson's discussion of the attempts of the PUMC-affiliated First National Midwifery School to retrain old-style midwives during the late 1920s and early 1930s in order to make their work more compatible with sanitary practices of childbirth. Tina Phillips Johnson, *Childbirth in Republican China: Delivering Modernity* (Lanham: Lexington Books, 2011), 93–102.

Chinese Forensics and the Global History of Science

Legal medicine, as it was established in China by Lin Ji and others during this period, must be understood as part of a global scientific enterprise. This is apparent, for example, if one looks through the ninth volume of the Rockefeller Foundation's *Methods and Problems of Medical Education* series, which surveyed the legal medicine institutes that were being built and planned in various countries.[50] Chinese medico-legal laboratories such as the department of legal medicine of the medical school of Beiping University and the significantly larger Research Institute of Legal Medicine (a central government forensic laboratory established in north-west Shanghai in 1932) followed institutional models that were widely accepted in many countries during this period.

We can find a valuable perspective for understanding the apparent transnational universality of scientific fields such as legal medicine in Timothy Lenoir's discussion of the ways in which scientific disciplines affect the practice of science and organize the relationship between science and society. Specifically, Lenoir argues that disciplines coordinate the local sites at which science is actually practiced, thus giving scientific knowledge and expertise a sense of universalism or unity that papers over, one might say, the highly localized configurations of skill, knowledge, and institutional support in which it is carried out on a day-to-day basis.[51] A similar point could be made when the modern history of science is viewed from a global perspective: scientific disciplines operate simultaneously at national and transnational scales of activity, established in support of the scientific communities of particular countries yet inseparable from broader networks of training and multi-lingual knowledge exchanges that cross national boundaries.[52] That one can even speak of a "world history of legal medicine and sciences," the title of the historian Jia Jingtao's magisterial overview of the world history of forensic practices, gives a sense of the important intellectual work that disciplines play as abstract divisions of human knowledge that allow us to imagine and reconcile the consequential but ambiguous relations between the national and the global in the transnational sciences of the twentieth century.[53]

---

[50] The Rockefeller Foundation, Division of Medical Education, *Methods and Problems of Medical Education (Ninth Series)* (New York: The Rockefeller Foundation, 1928).

[51] Timothy Lenoir, *Instituting Science: The Cultural Production of Scientific Disciplines* (Stanford: Stanford University Press, 1997), 45–51.

[52] These dynamics are illustrated vividly in Grace Shen's account of the ways in which early twentieth-century Chinese geologists positioned themselves vis-à-vis the scientific communities of other countries. Grace Yen Shen, *Unearthing the Nation: Modern Geology and Nationalism in Republican China* (Chicago: The University of Chicago Press, 2014), 73–107.

[53] Jia, *Shijie fayixue yu fakexue shi.*

Thus, as suggested by Lenoir's analysis, scientific disciplines provide conditions in which actors working in different global contexts can claim a coherent professional identity and compatible epistemological assumptions despite very real variations in the social, cultural, and historical contexts in which science is actually practiced.

That scientific practice must be understood in local contexts is readily demonstrated in a significant body of historical and ethnographic scholarship on the forensic sciences, which has revealed different ways in which scientists' relationships with legal professionals and the law shape the possibilities for them to attain authority in court and other societal arenas, and even the ways in which they handle and interpret physical evidence in the morgue or laboratory.[54] In early twentieth-century China, medico-legal experts' involvement in homicide cases was contingent on the decision of judicial officials to forego their own examination of the body or other evidence, as conducted by inspection clerks under their supervision, and send it to the laboratory instead. The practice of science in China's first forensic laboratories was shaped in different ways by officials of the judiciary given that in many cases they were the ones who collected the evidence and sent it for examination along with questions about its larger significance for the case at hand. As such, the work that officials and inspection clerks performed as intermediaries between law and science must figure prominently in any telling of the history of legal medicine in China during this period.[55] By implication, China's first legal medicine experts encountered institutional relationships and politics of knowledge that were unique to China's long history of bureaucratically organized forensic practices and that were not commonly shared by legal medicine experts and institutions in other countries during this same period.

Modern scientific disciplines such as legal medicine were instituted in early twentieth-century China against the backdrop of an expansive world of early modern natural knowledge and technical expertise of which the forensic examination techniques examined in this book constituted one

---

[54] Gary Edmond, "The Law-Set: The Legal-Scientific Production of Medical Propriety," *Science, Technology, & Human Values* 26, no. 2 (2001): 191–226; Timmermans, *Postmortem*, 113–56; Michael Lynch, Simon A. Cole, Ruth McNally, and Kathleen Jordan, *Truth Machine: The Contentious History of DNA Fingerprinting* (Chicago: The University of Chicago Press, 2008).

[55] For the roles that "non-experts" play more generally as intermediaries and participants in different fields of science, see Susan Leigh Star and James R. Griesemer, "Institutional Ecology, 'Translations' and Boundary Objects: Amateurs and Professionals in Berkeley's Museum of Vertebrate Zoology, 1907–1939," *Social Studies of Science* 19 (1989): 387–420; Fa-ti Fan, *British Naturalists in Qing China: Science, Empire, and Cultural Encounter* (Cambridge: Harvard University Press, 2004); Sigrid Schmalzer, *The People's Peking Man: Popular Science and Human Identity in Twentieth-Century China* (Chicago: The University of Chicago Press, 2008).

small corner. Indeed, much as David Arnold notes for nineteenth-century India, China was not a "scientific and technological *tabula rasa*" at the moment when new forms of modern science, medicine, and technology were introduced and established, a factor that must be taken into account when understanding the history of these new knowledge-practices.[56] The early twentieth century saw unprecedented challenges to old fields of Chinese knowledge as well as new kinds of interactions between older fields of knowledge and expertise and those introduced from other countries, especially Japan and the West. Generally speaking, the outcome of these encounters was a broad "displacement," as Benjamin Elman has characterized it, of old fields of knowledge and expertise in favor of those based on Japanese and Euro-American science and modern professional disciplines.[57] This outcome was the result of momentous shifts that included a general devaluing of the scholarly knowledge of the late imperial literati elite, new state-sponsored educational institutions that valorized modern fields of science and expertise, and, in the decades after the fall of the Qing, the rising influence of professional academics and scientists who established intellectual authority over myriad areas of natural and social inquiry. One suspects that in many cases the only possibility for older bodies of late imperial knowledge to "survive" in these conditions was for twentieth-century professionals or intellectuals to promote their own syntheses of old and new or to write about older practices when narrating the history of this or that modern discipline in China.[58]

The case of Chinese medicine is, of course, an important example of an old area of knowledge and expertise that underwent institutional, intellectual, and technological transformation over the course of the twentieth century, eventually giving rise to a deeply pluralistic world of knowledge and techniques that has been a component of the official healthcare system of the People's Republic of China since the 1950s.[59] The half-century that

---

[56] David Arnold, *Science, Technology and Medicine in Colonial India* (Cambridge: Cambridge University Press, 2004), 8–9. For recent scholarship that attempts to reconstruct the intellectual, social, cultural, and political contexts of different areas of technical knowledge and natural inquiry in late imperial China, see, for example, Benjamin A. Elman, *On Their Own Terms: Science in China, 1550–1900* (Cambridge: Harvard University Press, 2005); Carla Nappi, *The Monkey and the Inkpot: Natural History and Its Transformations in Early Modern China* (Cambridge: Harvard University Press, 2009); Dagmar Schäfer, *The Crafting of the 10,000 Things: Knowledge and Technology in Seventeenth-Century China* (Chicago: The University of Chicago Press, 2011).

[57] Elman, *On Their Own Terms*, 396–421.

[58] Jiri Hudecek, *Reviving Ancient Chinese Mathematics: Mathematics, History and Politics in the Work of Wu Wen-Tsun* (Abingdon: Routledge, 2014).

[59] Judith Farquhar, *Knowing Practice: The Clinical Encounter of Chinese Medicine* (Boulder: Westview Press, 1994); Volker Scheid, *Chinese Medicine in Contemporary China: Plurality and Synthesis* (Durham: Duke University Press, 2002); Kim Taylor, *Chinese Medicine in*

preceded the formation of Traditional Chinese Medicine represents, in many ways, an inversion of the history of forensics. Unlike healing knowledge and expertise in late imperial times, forensic knowledge had always been tightly regulated by the state and standardized by design. By contrast, proponents of Chinese medicine worked hard during the early twentieth century to demonstrate the relevance of their knowledge to the state and to obtain protection against the threat of unfavorable regulation if not abolishment.[60] The "Chinese medicine" that exists in China today is the product of long-term processes of formalization and standardization at the level of institutions, texts, and techniques, all of which has made this knowledge compatible with modern structures of professional training and official regulation as well as epistemological standards associated with the modern sciences.[61] The case of forensics again presents a contrast. From the outset, the Qing state's regime of forensic knowledge was compatible with the bureaucratic needs of the modern judiciary that was put into place during the first few decades of the twentieth century, an institution that in essence gave this knowledge an officially protected status and position. The close and long-standing relationship between late imperial forensic knowledge and the state is thus essential for understanding the trajectory of its modern history.

The case of forensics in Republican China stands out as a unique example of the possibilities for fields of late imperial knowledge and expertise to survive autonomously, one might say, of the new institutions of modern Western science. The forensic practices of the Qing state were not truly "displaced" until decades later with the unprecedented successes of medico-legal institution-building and training under the People's Republic of China. A simultaneous process of reimagining has taken place since 1949 in which the *Washing Away of Wrongs* has come to be identified not as an Other to Western legal medicine, a common

Early Communist China, 1945–63: A Medicine of Revolution (Abingdon: RoutledgeCurzon, 2005).

[60] Ralph C. Croizier, *Traditional Medicine in Modern China: Science, Nationalism, and the Tensions of Cultural Change* (Cambridge: Harvard University Press, 1968), 132–45; Sean Hsiang-lin Lei, *Neither Donkey nor Horse: Medicine in the Struggle over China's Modernity* (Chicago: The University of Chicago Press, 2014), 97–119, 141–166; Xu, *Chinese Professionals and the Republican State*, 192–3, 204–7.

[61] For more on the modern standardization of Chinese medicine, see Elisabeth Hsu, *The Transmission of Chinese Medicine* (Cambridge: Cambridge University Press, 1999), 128–67, as well as Scheid, *Chinese Medicine in Contemporary China*, 65–106. For a broader perspective on this process that explores parallels with Western medical professionals' attempts to standardize their own knowledge in early twentieth-century China, see David Luesink, "State Power, Governmentality, and the (Mis)remembrance of Chinese Medicine," in *Historical Epistemology and the Making of Modern Chinese Medicine*, ed. Howard Chiang (Manchester: Manchester University Press, 2015).

understanding during the Republican period, but rather as an example of China's precocious advances in this field of modern science, one that preceded comparable works of legal medicine in Europe by centuries. Likewise, Song Ci, author of the original *Collected Writings*, has been reimagined as a pre-modern "medico-legal expert" (*fayixuejia*) and progenitor of this modern scientific discipline in China and even globally, an assumption that elides the distinctive forms of judicially oriented, not medically oriented, forensic knowledge with which his text was long associated.[62]

By revisiting a period in which the modern discipline of legal medicine struggled with this older forensic tradition for institutional authority and legitimacy, at times on highly unfavorable terms, we gain an understanding of the interaction between two forensic regimes that construed the examination of bodies and things in very different ways. We also gain insight into the process through which Chinese forensic scientists and others have come to view the *Washing Away of Wrongs* as a kind of pre-modern stage in the inevitable rise of legal medicine in China, an intellectual move that adds another "national" chapter to the discipline-history of modern forensic science even as it simplifies the complex and contested nature of these forensic regimes' early twentieth-century interactions.

Forensics and the History of Death in China

Investigating homicide often took a toll on the dead body. Bodies might lie exposed in public while police waited for officials of the procuracy to arrive at the scene. Forensic examinations by this office were often conducted in the open, with the surface of the body thoroughly inspected and probed in front of relatives of the deceased, police officials, and urban spectators. Bodies that had decomposed might be subjected to a "steaming examination" (*zhengjian*), a procedure that involved disarticulating the bones and then steaming them along with wine, vinegar, and distiller's grains in order to make traces of wounds visible on the bone surfaces. In cases of suspected poisoning, inspection clerks might insert a silver hairpin or probe into the dead body, a test that would provide proof of poisoning if a tell-tale dark discoloration appeared. When this test was not used and officials decided to make use of the toxicological testing services of one of the new medico-legal laboratories in Beijing or Shanghai, internal organs might be extracted from the body, placed in

---

[62] Song Daren, "Weida fayixuejia Song Ci zhuanlüe" [A brief biography of the great medico-legal expert Song Ci], *Yixue shi yu baojian zuzhi* 2 (1957): 116.

a container, and mailed for testing. In other cases in which physicians became involved, a body might be autopsied and the internal organs inspected for signs of the trauma or disease pathology that led to death.

How we should interpret this history of exposed and damaged bodies is one of the main questions of this book. The first few decades of the twentieth century saw significant shifts in the institutional management of dead bodies in China, in the sources of practical knowledge with which they were investigated in medical contexts, and in normative expectations about what should happen to the body after death.[63] At the same time, Chinese interest in the dead body as an object of legal concern and technical knowledge, especially in forensics, had long preceded the modern period. The twentieth century saw, rather, a reorganization of existing concerns, institutions, and methods, some of which had existed for centuries. One of the challenges of approaching this history is grappling with a long-standing and diffuse discourse, examples of which can be found in the nineteenth century if not earlier, that has posited a culturally rooted tendency of the Chinese to revere and protect the dead body, an attitude that purportedly hinders the pursuit of more practical or utilitarian goals where the dead are concerned. Thus, for example, we find H.B. Morse (1855–1934) explaining the riot that erupted following the attempts of authorities in Shanghai's French Concession to build two roads through a portion of the cemetery of Ningbo guild members as a result of the fact that of "all the duties of a provincial club, that of providing the mortuary and the cemetery is held to be the most important; and any action affecting either adversely is most deeply resented, *respect for the dead and the desire to rest in the home soil being most deeply implanted in the Chinese character* [italics added]."[64] In his own account of the 1874 riot, F.L. Hawks Pott concluded,

We have dwelt at length upon this first riot, as it is a good example of difficulty arising between two peoples on account of their different points of view. To the Westerner, with his desire for that which is useful, it seemed absurd that the construction of a road should be held up by the unwillingness of the Chinese to remove some graves. To the Chinese it seemed that the Westerner was wanting in respect for the dead, the strongest cult in China.[65]

In a similar vein, those who promoted the progress of Western scientific medicine in Republican China could be found invoking the country's "great reverence for the dead" to explain the fact that human dissection, while attempted at times, had never been foundational to Chinese

---

[63] Luesink, "Dissecting Modernity."

[64] Hosea Ballou Morse, *The Gilds of China* (London: Longmans, Green and Co., 1909), 47.

[65] F.L. Hawks Pott, *A Short History of Shanghai* (Shanghai: Kelly & Walsh, Ltd., 1928), 98.

physicians' understandings of the body (similarly, Chinese medical reformers might invoke an idea of indigenous *fengsu*, or "customs," to explain
the difficulties that medical schools faced in procuring bodies for
dissection).[66] The problem with these ways of characterizing Chinese
approaches to the dead body is not in the claim that Chinese had "respect
for the dead," a recognizable sentiment in many times and places, or that,
as a large body of scholarship has convincingly shown, late imperial death
ritual and Cheng-Zhu orthodoxy shaped normative expectations about
how the living should respond to death and how the dead body should be
treated.[67] The problem, rather, is the lack of agency that cultural-
essentialist explanations imply – the idea that there is a singular cultural
mindset that informs if not constrains the behavior of all. Such an
assumption makes it harder to recognize the ways in which Chinese
approaches to the dead body might have differed across place, time, or
even class, and the ways in which particular engagements with the dead
were the result of individuals' pragmatic negotiations of different imperatives or concerns, whether legal, ritual, or otherwise.[68]

In this sense, the question of what constituted a "proper" way of
handling the dead body in late imperial and early twentieth-century
China is a more complex question than it might seem. In a study of
eighteenth- and nineteenth-century legal cases involving the crime of
"uncovering graves" (*fazhong*), Jeff Snyder-Reinke has shown that people
opened graves and disturbed bodies for a number of reasons, including to
extort others, steal valuables, reclaim land for farming, protect the geo-
mantic properties of a burial ground, or even procure body parts for
medicinal use.[69] In her study of late imperial litigation masters, Melissa
Macauley has shown that litigants might use a dead body to push forward
a false accusation of murder and advance their own interests in the legal
arena, a very practical use of the dead that was not informed by ritual

---

[66] Edward H. Hume, "The Contributions of China to the Science and Art of Medicine,"
Science 59, no. 1529 (1924): 348.

[67] For general perspectives on death ritual in late imperial China, see James L. Watson and
Evelyn S. Rawski, ed., *Death Ritual in Late Imperial and Modern China* (Berkeley:
University of California Press, 1988); Donald S. Sutton, "Death Rites and Chinese
Culture: Standardization and Variation in Ming and Qing Times," *Modern China* 33,
no. 1 (2007): 125–53.

[68] For the diversity of interests and motives that the living brought to their interactions with
dead bodies in late imperial China, see Jeff Snyder-Reinke, "Afterlives of the Dead:
Uncovering Graves and Mishandling Corpses in Nineteenth-Century China," *Frontiers
of History in China* 11, no. 1 (2016): 1–21. Also see Tobie Meyer-Fong's study of the
varied elite and popular responses to mass death as experienced during the Taiping civil
war. Tobie Meyer-Fong, *What Remains: Coming to Terms with Civil War in 19th Century
China* (Stanford: Stanford University Press, 2013), 99–174.

[69] Snyder-Reinke, "Afterlives of the Dead."

propriety.[70] In his own study of false accusation cases from late nineteenth-century Ba county, Sichuan, Quinn Javers similarly notes that "[death] was an unexpected opportunity for an economic windfall."[71] We find a no less complex set of motives in the late imperial state's deployment of judicial dismemberment (*lingchi*), a punishment that, as Timothy Brook, Jérôme Bourgon, and Gregory Blue have argued, was considered to be so severe because it caused more damage to the dead body than did decapitation or strangulation.[72] While, as the authors suggest, the history of this punishment provides a window onto popular ideas about the importance of maintaining the "somatic integrity" of the corpse for the sake of the deceased person in the afterlife, officials' use of this punishment indicates as well that the state routinely accepted the physical destruction of the body in order to further the legal and political interests at stake in the administration of justice.

The history of homicide investigation also provides insight into the diversity of ways in which Chinese approached the dead body as well as the ways in which ritual imperatives were reconciled with other concerns – in the case of forensics, the epistemological drive for knowledge about what caused a death and the legal drive to serve justice. The practical imperatives of homicide investigation sat in a complex, at times ambivalent, relationship with other dimensions of early twentieth-century China's changing culture of death as it was experienced in particular local contexts. In Beijing, an investigation by police and procurators might delay burial or otherwise hinder a mourning family's initiation of funeral rites, but it could also facilitate the efficient burial of the unclaimed bodies of the unidentified or destitute when these fell under the gaze of the urban state. The procuracy's examination of a body, or even a physician's autopsy, could cause a body to be exposed and even physically damaged. The same procedures, however, could be used to solve a murder, document wrongdoing, and pursue justice – desirable outcomes for the relatives of the deceased. From the perspective of forensic examiners – whether the procuracy's forensic personnel or legal medicine experts such as Lin Ji – the questions of what happened to bodies, how soon they were buried, and in what condition were intrinsically related to their own capacity to demonstrate their forensic skills and

---

[70] Melissa Macauley, *Social Power and Legal Culture: Litigation Masters in Late Imperial China* (Stanford: Stanford University Press, 1998), 197–206.

[71] Quinn Javers, "The Logic of Lies: False Accusation and Legal Culture in Late Qing Sichuan," *Late Imperial China* 35, no. 2 (2014): 31.

[72] Timothy Brook, Jérôme Bourgon, and Gregory Blue, *Death by a Thousand Cuts* (Cambridge: Harvard University Press, 2008), 13–4.

assert authority – concerns that were different from those of the family of the deceased or others.

In all of these ways, forensic encounters with the dead body, however physically damaging they might be, should not simply be viewed as opposing or challenging normative expectations surrounding the "proper" handling of the dead as defined by ritual or otherwise. Rather, the process of forensic investigation was itself productive of social meaning in multiple ways.[73] From this perspective, law and science were as important as funeral ritual for the ways in which people in early twentieth-century China negotiated the bonds between living and dead, confirmed the identity and social value of the deceased, and imagined and acted upon the society in which they lived.

---

[73] For perspectives on the social and affective meanings of human remains in various historical contexts, see the collection of essays in the December 2010 issue of *Journal of Material Culture*, especially Elizabeth Hallam, "Articulating Bones: An Epilogue," *Journal of Material Culture* 15, no. 4 (2010): 465–92. Zoë Crossland has argued that in forensic anthropology and related human sciences, the authority of (scientific) evidentiary claims made about bodies and body parts derive in part from a powerful symbolic association between these physical "things" and the social identity of the deceased person. By implication, even forensic practices that seem to physically damage the dead body have a stake in validating, not ignoring, the identity of the deceased as a formerly living person. Zoë Crossland, "Of Clues and Signs: The Dead Body and Its Evidential Traces," *American Anthropologist* 111, no. 1 (2009): 69–80.

1   Suspicious Deaths and City Life in Republican Beijing

Opening to the seventh page of the January 9, 1928, issue of *Morning Post*, a reader of this esteemed Beijing newspaper would have found the story of an unnamed man who was horrifically mangled under the wheels of a train on the Beijing-Suiyuan railway line.[1] The incident had occurred on the previous day, just outside of the walls enclosing the southwestern quadrant of the Inner City. This line, one of several that rounded the perimeter of the walled city, began operation between Beijing and Kalgan in 1909 and provided connections to Mukden- and Hankou-bound trains in Fengtai, to the southwest of Beijing.[2] The man had been crossing the tracks, the piece reported, when "suddenly he saw the train speeding toward him. Bewildered and panicking, the man was paralyzed with fright as he hesitated on the tracks not knowing what to do." His legs failing him, this unfortunate person was run over by the train, which left in its wake a horrifically mutilated body splayed across the tracks. As was typical in the newspaper's reporting of deaths, the state of the corpse was described in detail: "The space from the man's lower chest to left upper arm had been run over by the wheels, smashing bone and mashing flesh and cutting him in two. His internal organs had spilled out and were covering the ground. Bright red blood splashed all over the tracks." The condition of the corpse was "too awful to look at," but clearly not too awful for a newspaper readership accustomed to daily coverage of murders, suicides, and the machine-produced horrors of an industrializing urban landscape. Members of the Beijing police soon arrived and, as was customary following violent or otherwise unexpected deaths, summoned members of the Beijing procuracy to the scene. Following a brief investigation into the circumstances of the death and an examination of the corpse, officials of this office confirmed that no crime had been committed and released the body for burial.

---

[1] *Chenbao* (CB) January 9, 1928, 7.
[2] H.G.W. Woodhead, *The China Year Book 1929–30* (Tientsin: The Tientsin Press, 1930), 403–4.

More than simply a sensational account of tragic and accidental violence, this *Morning Post* news report revealed a basic, unavoidable fact about death in Beijing: no dead body was permitted to be buried without having undergone some level of official investigation. In most cases, this simply meant that relatives of the person who died had to report the death to the police and obtain a permit that was needed to bring the body through the city gates on the way to a burial ground. For those who died alone and without relatives or others to identify the body, who died violently, or whose bodies were discovered in public spaces, a very different set of procedures was carried out by police and procurators; as in this case, this process revolved around a public examination of the body, a procedure that was covered in the newspapers. This chapter examines the ways in which this daily search for homicides, carried out through police procedures for identifying suspicious or unexpected deaths in need of further investigation by the procuracy, shaped the modern culture of death in the city.

Examining the claims made by city authorities over the dead in Beijing can reveal some of the ways in which the social, economic, and cultural changes associated with modernity transformed the meanings of death and the practical management of dead bodies in China. This was a period in which Chinese responses to death were, broadly speaking, brought more closely into line with those of industrialized countries in the West and Japan.[3] Death in Beijing and other urban areas became the object of a multiplicity of specialized disciplines: experts in public hygiene, pathology, sociology, and other fields all made new claims regarding their ability to explain death individually and in aggregate, in the process making death in China visible through new, transnationally constituted fields of knowledge and expertise.[4] Yet, these new ways of establishing knowledge of mortality and managing the dead body were established within the context of older expectations about what should happen to the body following death. In Beijing, these expectations were drawn from the existing world of funeral ritual and the legacies of urban policing in the

---

[3] For the modern transformation of death in Western societies, see Philippe Ariès, *The Hour of Our Death*, translated by Helen Weaver (New York: Alfred A. Knopf, 1981); Lindsay Prior, *The Social Organization of Death: Medical Discourse and Social Practices in Belfast* (New York: St. Martin's Press, 1989); Thomas A. Kselman, *Death and the Afterlife in Modern France* (Princeton: Princeton University Press, 1993). For the case of Japan, see Andrew Bernstein, *Modern Passings: Death Rites, Politics, and Social Change in Imperial Japan* (Honolulu: University of Hawai'i Press, 2006).

[4] For the role that scientific disciplines play in making death visible in modern societies, see Prior, *The Social Organization of Death*. For the visibility of suicide in social surveys and journalistic reporting in Republican China, see Peter J. Carroll, "Fate-Bound Mandarin Ducks: Newspaper Coverage of the 'Fashion' for Suicide in 1931 Suzhou." *Twentieth-Century China* 31, no. 2 (2006): 75–9.

former Qing imperial capital, as well as the centralized system of forensic body examination that had been implemented under the fallen empire.

Thus, much as Madeleine Yue Dong has argued for other aspects of economic, social, and cultural life in Republican Beijing, official and societal responses to death were characterized, on many levels, by the "recycling" of old rituals, policing practices, and knowledge, all of which were given new utility and meaning in the modern urban order that was emerging during this period.[5] This melding of old and new manifested in numerous ways, ranging from the modern urban state's use of ritual specialists to certify that a death had not involved suspicious circumstances to its reliance on Beijing's majestic city gates as a control point for supervising the passage of dead bodies out of the city. The modernity of death in Republican Beijing thus must be sought in very local configurations of institutions, rituals, spaces, and knowledge, as well as the new, permeating forces of industrial print capitalism, which created a particular vision of urban life and death that was disseminated on a daily basis in the newspapers.

Funeral and Burial Practice in Early Twentieth-Century Beijing

By the mid-1920s, Beijing was a city of over a million people. The city itself comprised Inner and Outer Cities, two large walled areas with a combined area of approximately 25 square miles, as well as four large suburban districts that surrounded the walled city on all sides with an area of approximately 253 square miles.[6] The Metropolitan Police Board (Jingshi jingchating), Beijing's modern police force, divided the Inner and Outer Cities into districts, each of which handled a range of administrative and quasi-judicial functions in the diverse areas that they controlled (Figure 1.1). Beyond investigating crime and maintaining order on the streets, district police collected census information, enforced hygiene regulations, suppressed sex-themed media and other purportedly deviant behaviors, and provided soup kitchens and other forms of relief for the urban poor while simultaneously shuttling them across an evolving network of coercive workhouses and detention facilities.[7] Within each

---

[5] Madeleine Yue Dong, *Republican Beijing: The City and Its Histories* (Berkeley: University of California Press, 2003), 10–12, 306–7.

[6] Han Guanghui, *Beijing lishi renkou dili* [Historical demographic geography of Beijing] (Beijing: Beijing daxue chubanshe, 1996), 133; Sidney D. Gamble, *Peking: A Social Survey* (New York: George H. Doran Company, 1921), 413; Robert Moore Duncan, *Peiping Municipality and the Diplomatic Quarter* (Tientsin: Peiyang Press, Ltd., 1933), 41.

[7] Gamble, *Peking*; Ng, *Legal Transplantation in Early Twentieth-Century China*, 30–52; Strand, *Rickshaw Beijing*, 83–9; Hugh L. Shapiro, "The View from a Chinese Asylum: Defining Madness in 1930s Peking" (PhD dissertation, Harvard University, 1995),

GEOGRAPHY

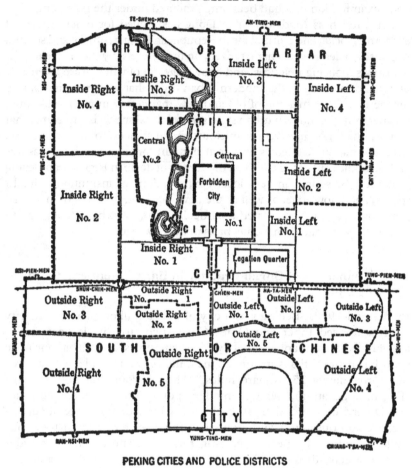

PEKING CITIES AND POLICE DISTRICTS

1.1 Map of Beijing police districts, early 1920s.

police district, urban spaces were policed by officers located in numerous sub-stations as well as those who patrolled the streets. The original twenty

82–113; Yamin Xu, "Wicked Citizens and the Social Origins of China's Modern Authoritarian State: Civil Strife and Political Control in Republican Beiping, 1928–1937" (PhD dissertation, University of California, Berkeley, 2002); Janet Y. Chen, *Guilty of Indigence: The Urban Poor in China, 1900–1953* (Princeton: Princeton University Press, 2012), 62–70, 96–107; Y. Yvon Wang, "Whorish Representation: Pornography, Media, and Modernity in Fin-de-siècle Beijing," *Modern China* 40, no. 4 (2014): 351–92.

districts of the Beijing police were consolidated into eleven following the reorganization of this agency into a Bureau of Public Safety in 1928 after the city came under the control of the Nationalist party-state, which reorganized the city administration according to its new laws on municipal government.[8] Throughout the Republican period, Beijing's districts varied significantly in population size and density, commercial activity, infrastructure, and patterns of crime and policing. There were significant differences, for example, between the busy and densely packed markets that stretched south from Qian Gate to Tianqiao and the less populous edges of the Outer City, a much less developed area of Beijing that was covered in burial grounds.[9]

Generally, early twentieth-century Beijing was not a city in which people died in hospitals, nor was it a city in which popular understandings of morbidity and mortality were defined by physicians or institutions of Western scientific medicine, the medicine of anatomic pathology and germ theory. This reflected a general paucity of institutions of Western medicine as well as the very real economic constraints that informed city dwellers' decisions about healthcare in a city for which widespread poverty shaped many aspects of urban life.[10] Nor was Beijing a city in which city dwellers' responses to death were coordinated by commercial funeral homes (*binyiguan*), even though funerary services and paraphernalia could be obtained commercially and at a wide range of prices through funeral specialists, coffin shops, and fee-based coffin-storage facilities and cemeteries.[11] Native-place associations (*huiguan*) and occupational

[8] Duncan, *Peiping Municipality and the Diplomatic Quarter*; Mingzheng Shi, "Beijing transforms: Urban infrastructure, public works, and social change in the Chinese capital, 1900–1928" (PhD dissertation, Columbia University, 1993), 25–87. For the organization of police institutions, see Han Yanlong and Su Yigong, *Zhongguo jindai jingcha shi* [A history of modern policing in China] (Beijing: Shehui kexue wenxian chubanshe, 2000), 335–52; Gamble, *Peking*, 75–80; Gong Yibing, "Beijing jindai jingcha zhidu zhi quhua yanjiu" [Research on the district divisions of the modern police in Beijing], *Beijing shehui kexue* no. 4 (2004): 104–14.

[9] For the development and importance of Tianqiao market district, see Dong, *Republican Beijing*, 172–207. For district-level population figures as of the late 1910s, see Gamble, *Peking*, 412–3. For differences in crime rates, see Ng, *Legal Transplantation in Early Twentieth-Century China*, 98–111.

[10] In late-1910s Beijing, for example, the ratio of physicians of Western medicine to population was approximately 1:7,400 (if physicians of Chinese medicine are counted as well, the figure jumps to 1:739). There were several dozen hospitals in early Republican Beijing and these varied considerably in institutional scale, sources of funding, and national and religious affiliation. Gamble, *Peking*, 118–9, 412.

[11] For funeral homes in Republican Shanghai, see Huwy-min Lucia Liu, "Dying Socialist in Capitalist Shanghai: Ritual, Governance, and Subject Formation in Urban China's Modern Funeral Industry" (PhD dissertation, Boston University, 2015), 37–8, 53–60. For Beijing funeral expenses, see Sidney D. Gamble, *How Chinese Families Live in Peiping* (New York: Funk & Wagnalls Company, 1933), 211–9, 231–41.

guilds, mutual-aid associations that provided for members' funerals, and even district police, who routinely encoffined and buried the bodies of the poor, provided some of the organized infrastructure through which the living responded to death.[12] Most commonly, though, it was the family that served as the most important social milieu in which city dwellers made sense of death and managed its immediate aftermath. The home is thus an important starting point for thinking about how state and society responded to death in the urban context of 1920s and 1930s Beijing.

Most immediately, the home of the deceased provided the setting for a complex set of funeral rituals. Prior to bringing the coffin out of the home, members of the family might wash, shroud, and encoffin the corpse, employ a ritual specialist known as a "yinyang master" (*yinyang sheng, yinyang xiansheng*) to establish a timeline for the rest of the funeral arrangements and interment, and seek out Buddhist monks or other kinds of priests to perform an array of rituals meant to ensure a good afterlife for the deceased and protect the family and other mourners from the inauspicious dimensions of death.[13] As Susan Naquin has noted, there were many variations in the diverse funerary repertoires that were used in north China during this period.[14] In a city with a significant Muslim population, for example, Beijing funerals could involve the participation of religious officiants from this community and a correspondingly eclectic ritual repertoire.[15] The economic means of a family was also a basic factor in determining the appearance and form of a funeral. Everything from the

---

[12] For the role of native-place lodges and guilds in managing funerary arrangements, maintaining cemeteries, and organizing the storage and transport of coffins, see Richard Belsky, *Localities at the Center: Native Place, Space, and Power in Late Imperial Beijing* (Cambridge: Published by the Harvard University Asia Center. Distributed by Harvard University Press, 2005), 95–6, 121–6, and 153–4, as well as the materials collected in Saeki Yūichi, Tanaka Issei, Hamashita Takeshi, and Ueda Shin, eds., *Niida Noboru hakushi shū Pekin kōshō girudo shiryōshū* [A collection of materials on industrial and commercial guilds of Beijing, compiled by Dr. Niida Noboru] (Tokyo: Tōkyō daigaku Tōyō bunka kenkyūjo Tōyōgaku bunken sentā, 1975–83). For the case of the Old Ladies' Home, a welfare institution supported by private foreign funds that paid for the encoffining and burial of the women who lived in its facilities, see Gamble, *Peking*, 301. For mutual-aid societies, see Li Jiarui, *Beiping fengsu leizheng* [A categorized collection of Beiping customs] (Shanghai: Shangwu yinshuguan, 1937), 147.

[13] For an overview of death customs in late-nineteenth- and early twentieth-century Zhili/Hebei based on local gazetteers, see Susan Naquin, "Funerals in North China: Uniformity and Variation," in *Death Ritual in Late Imperial and Modern China*, ed. James L. Watson and Evelyn S. Rawski (Berkeley: University of California Press, 1988). Descriptions of Beijing funeral customs can also be found in Li, *Beiping fengsu leizheng*, 129–32 and Chang Renchun, *Lao Beijing de fengsu* [Customs of old Beijing] (Beijing: Beijing yanshan chubanshe, 1996).

[14] Naquin, "Funerals in North China," 46–53.

[15] Gamble, *Peking*, 372–3; Naquin, "Funerals in North China," 50; Gamble, *How Chinese Families Live in Peiping*, 211–2, 215, 231–2.

type of coffin that was purchased to the ritual specialists who were employed to the composition and appearance of the funeral procession could range widely in cost. The total expenditures associated with fourteen funerals documented by Sidney Gamble, for example, ranged from 1.82 to 744.85 yuan, with 149.68 yuan as the average expenditure.[16] The size and cost of funerals for public figures or those whose deaths had other public significance could be staggeringly disproportionate. It was reported in the newspaper *World Daily*, for instance, that the funeral arrangements for Lin Kongtang, a 23 *sui* agriculture student who was killed during the massacre of protestors that took place on March 18, 1926, were initially budgeted at 1,000 yuan: 400 yuan for the coffin and funeral garments, 100 yuan for the memorial meeting, and the rest of the money for burial and the erection of a commemorative stele.[17]

One of the most visible elements of funeral practice in the city was the funeral procession, a public ritual event in which a mourning family transported the encoffined body through the city to a burial ground, accompanied (if economic resources allowed) by a spectacular train of musicians, priests, procession attendants, and even representatives of urban associations (Figure 1.2).[18] These processions competed for attention and space with the other people, activities, and, by the 1920s, machines that also occupied urban spaces. In August 1930, for example, a streetcar hit several members of a large, slow-moving procession near Dongsi Arch, causing the coffin to fall to the ground (The irate son-in-mourning refused to move the coffin off the tracks until officials of the procuracy could inspect the body and confirm that it had not been damaged).[19] Such processions could take on social and political meaning beyond that of a family's mournful transportation of the encoffined body to burial. Funeral processions carried out for the students and others who were killed in the March Eighteenth massacre, for example, featured symbols and actions that constructed the victims of the killing as patriotic martyrs (*lieshi*) whose wrongful deaths demanded a political response. The funeral procession for Yao Zongxian, a 21 *sui* student from Yunnan who was killed in the massacre, proceeded from the art school at which Yao had studied, south through Xuanwu Gate, and then west toward Guang'an Gate, finally ending at Zhaogong Shrine (Zhaogong ci),

---

[16] Naquin, "Funerals in North China," 48–9; Gamble, *How Chinese Families Live in Peiping*, 231–41.

[17] SJRB 29 May 1926, 7. In this book, ages are given in *sui*, which is a Chinese way of calculating age that is counted by adding an additional year of age on the lunar New Year following birth, and every one thereafter.

[18] Li, *Beiping fengsu leizheng*, 127–9, 147; Gamble, *How Chinese Families Live in Peiping*, 213–5.

[19] SJRB August 13, 1930, 7.

1.2 Beijing funeral procession, late 1910s.

a Yunnan native-place lodge. As reported in *Morning Post*, those accompanying the procession shouted slogans such as "Vow to kill the national traitors!," "Avenge the martyrs!," and "Long live the 'citizens' revolution' (*guomin geming*)!"[20] Such processions could take on other meanings and be used in other ways as well. When a telegraph bureau employee was fatally injured by a streetcar in early March 1927, it was newsworthy that relatives and coworkers used the funeral procession to press their demand that the streetcar company provide compensation to the victim's family.[21] Members of the procession walked the casket slowly along the streetcar tracks near Chongwen Gate, in the process blocking streetcar traffic in that part of the city.

[20] CB March 25, 1926, 6. Compare with the June 1926 funeral procession for 27 *sui* Tan Jijian, another victim of the killings, which reportedly included the carrying of a national flag (*guoqi*), followed by a band, a photograph of Tan, a wreath-carrying troupe composed of individuals from Tan's native Hunan, Tan's family, and various association representatives and other guests. SJRB June 5, 1926, 7; SJRB June 7, 1926, 7; SJRB June 8, 1926, 7.

[21] CB March 9, 1927, 6.

City dwellers in early twentieth-century Beijing might bury their dead in private family burial grounds or graveyards that were open to the public or affiliated with institutions such as native-place lodges and guilds, store the coffin in a temple for later burial, or transport the body out of the city for burial elsewhere. Ideally, interment took place in family burial grounds located beyond the walls of the city. During the 1930s, the suburban districts surrounding Beijing appear to have suffered an epidemic of grave robberies as gangs of looters stole countless valuables from such graves. The sites that were looted included family plots as well as the graves of famous individuals, such as the influential Manchu high official and military commander Ronglu (1836–1903), whose body was reportedly damaged when looters ransacked the grave.[22] The geography of organized burial grounds in the city was distinctive. A survey of cemeteries conducted in the mid-1930s found a total of 130 burial grounds, concentrated within the western and eastern edges of the Outer City and in the Southern suburban district, just to the south of the Outer City wall. A number of other burial grounds were located in the Western and Eastern suburbs.[23] This pattern was the result of long-term trends in the city's geography of burial that included the establishment of state-administered charitable cemeteries (*yizhong*) in the Outer City and at a number of points surrounding the city walls in the eighteenth century and the proliferation of cemeteries associated with native-place lodges in the Outer City during this same period and after.[24]

Upon the fall of the Qing, burial grounds in Beijing were owned by a wide range of individuals and organizations, including occupational guilds, native-place lodges, places of religious worship, and the city government itself. While a majority of public burial grounds surveyed in the 1930s did not charge fees, a number did charge 1–2 yuan per burial

---

[22] SJRB July 11, 1933, 8; SJRB October 28, 1933, 8; SJRB February 27, 1936, 6.

[23] The survey was completed in 1935 by the city government's Self-Government Affairs Supervisory Office (Zizhi shiwu jianlichu) and the Bureau of Social Affairs. Information was compiled into tables listing all temples storing coffins and burial grounds located under the jurisdictions of the fifteen self-government district offices of the city. BMA J2-7-117, 116–37.

[24] For the spatial distribution of burial grounds in mid-nineteenth century Beijing, see Dudgeon's brief description in *Medical Reports for the half year ended 30th September 1872; Forwarded by the surgeons to the customs at the treaty ports in China; Being no. 4 of the series, and forming the sixth part of the Customs Gazette for July–September 1872*. Published by order of the Inspector General of Customs (Shanghai: Printed at the Customs Press, 1873), 29–30. Also see Susan Naquin, *Peking: Temples and City Life, 1400–1900* (Berkeley: University of California Press, 2000), 647. For public cemeteries in Beijing during the Qing, see *Qinding Da Qing huidian shili* (Compiled by Kungang et al. [1899]. Reprinted in *Xuxiu siku quanshu*. Shanghai: Shanghai guji chubanshe, 1995), *juan* 1036, 812.397.

(an outlier at the high end was Wan'an Cemetery, located in the Fragrant Hills to the northwest of the city, which charged 70 yuan per plot).[25] These spaces served other functions beyond burial. In early June 1927, for example, a fire broke out at the cemetery of the Fujian-affiliated Quanjun native-place lodge when piles of finished matting and bamboo stored by an adjoining shop went up in flames.[26] In late December of the same year, it was announced that a soup kitchen would be held on the cemetery grounds of the Fujian-affiliated Puyang native-place lodge.[27] The public burial grounds that the city PSB maintained to the south of Yongding Gate, just beyond the walls of the Outer City, were used for executions.[28]

There were also many temples and other religious and secular establishments that stored coffins for weeks, months, or years. According to city authorities' mid-1930s survey, there were more than seventy temples and other facilities, many of which were associated with native-place lodges and guilds, which were distributed throughout the walled city and surrounding suburbs and stored 2,772 coffins.[29] The sites themselves ranged from temples containing a few coffins to larger facilities containing dozens, if not a hundred or more, with some storage sites tightly regulated and others containing old, seemingly abandoned coffins. Such services were especially important for Beijing's significant population of sojourners, individuals who might hope for their remains to be transported back to their native locality as well as others who lived in the city but were unable to conduct a funeral or bury the body immediately.[30] The Beijing police made use of this network of above-ground body storage sites: following the March Eighteenth massacre, the police decided to temporarily store the bodies of the victims that remained unclaimed at Xifang Temple, just outside of Dongzhi Gate, pending identification. Nevertheless, because the city gate had already closed for the day, the bodies were kept overnight at a Medicine King temple located inside of the walled city.[31] This is one example among many of the ways in which city authorities' engagements with the urban dead became entangled with city dwellers' existing death customs.

---

[25] BMA J2-7-117, 127–37.   [26] SJRB June 8, 1927, 7.
[27] SJRB December 23, 1927, 7.   [28] SJRB April 4, 1933, 8.
[29] "Table of survey results regarding stored coffins in temples located under each of Beiping Municipality's Self-Government Affairs district offices," 1935, BMA J2-7-117, 116–26; Chang, *Lao Beijing de fengsu*, 217–22.
[30] Chang, *Lao Beijing de fengsu*, 214–5.
[31] "Youguan sanyiba can'an sishang qingkuang de dang'an" [Archival materials on the situation of the dead and wounded of the March Eighteenth massacre], *Beijing dang'an shiliao*, no. 1 (1986): 22–6.

Burial Permits, City Gates, and Urban Surveillance

The Metropolitan Police Board, the police department that responded to the death of the man on the tracks of the Beijing-Suiyuan railway line in early January 1928, was the product of deep changes in conceptions and practices of governance that took place in China during the late Qing New Policies reforms. Following the disastrous aftermath of the Boxer uprising, the Qing court promoted a series of wide-ranging reforms, based significantly on Japanese models, that had the goal of creating a modern constitutional system that would strengthen the empire and make it competitive in the world of industrializing nation-states. The establishment of modern police institutions during this period grew out of a number of different factors, including decades of rising merchant and gentry managerial activism and authority and the new imperatives faced by the Qing to claim sovereignty internationally by demonstrating acceptance of Euro-Japanese political institutions and the related conceptions of domestic order and "civilization" with which they were connected.[32] In Beijing, it was out of the administrative vacuum that existed after the Boxer uprising – when non-Qing forces as well as ad hoc organizations arranged by local elites policed the city – that modern policing as implemented by Kawashima Naniwa (1865–1949), who oversaw the policing of the Japanese-occupied areas of Beijing during this period, emerged as a compelling new vision for reformers.[33] These reforms yielded policing institutions in Beijing that were consolidated by Yuan Shikai (1859–1916) in 1913 to form the Metropolitan Police Board, an institution that exercised unified authority over the Inner and Outer Cities in tandem with the Qing-era Gendarmerie, until the disbandment of the latter by militarist Feng Yuxiang (1882–1948) in 1924.[34]

---

[32] I.e. Frederic Wakeman, Jr., *Policing Shanghai 1927–1937* (Berkeley: University of California Press, 1995), 18–22; Kristin Stapleton, *Civilizing Chengdu: Chinese Urban Reform, 1895–1937* (Cambridge: Published by the Harvard University Asia Center and distributed by Harvard University Press, 2000); Tong Lam, "Policing the Imperial Nation: Sovereignty, International Law, and the Civilizing Mission in Late Qing China," *Comparative Studies in Society and History* 52, no. 4 (2010): 881–908. For a sophisticated treatment of the historical origins and intellectual politics of nineteenth- and early twentieth-century Western assumptions about the despotic nature of imperial China's legal and political institutions and the particular concepts of "sovereignty" and "civilization" on which these assumptions were based, see Li Chen, *Chinese Law in Imperial Eyes: Sovereignty, Justice, and Transcultural Politics* (New York: Columbia University Press, 2016).

[33] Han and Su, *Zhongguo jindai jingcha shi*, 82–92; Strand, *Rickshaw Beijing*, 67–8. For the evolution of administrative and policing institutions in the city during the Qing, see Naquin, *Peking*, 287–301; Alison Dray-Novey, "Spatial Order and Police in Imperial Beijing," *The Journal of Asian Studies* 52, no. 4 (1993): 885–922.

[34] Han and Su, *Zhongguo jindai jingcha shi*, 335–58; Alison J. Dray-Novey, "The Twilight of the Beijing Gendarmerie, 1900–1924," *Modern China* 33, no. 3 (2007): 349–76.

By the mid-1920s, the Beijing city police had put into place a multifaceted regime of surveillance over death in the city. District police, along with legal officials and hygiene authorities, maintained surveillance over the deaths of Beijing city dwellers for three broad purposes. One was to track deaths in order to maintain up-to-date records on the living arrangements and employment of district residents and the demographic composition of the urban population.[35] While the old Gendarmerie had collected information about individual residents as well as persons associated with businesses and temples, the investigations of the Republican police force surpassed these efforts in ambition and scope, all while bringing this work under the centralized authority of a single police agency.[36] The second concern was the collection of vital statistics for the surveillance of epidemic diseases and, more generally, to establish an understanding of urban patterns of morbidity and mortality. While investigations of death for purposes of public health had been carried out by police authorities in Beijing since the last decade of the Qing, the involvement of Peking Union Medical College (PUMC) in urban public health during the mid-1920s changed the nature of this work considerably by creating new institutional models and procedures that were adopted by the city government's own hygiene offices during the 1930s and 1940s.[37] Finally, police and procurators were concerned with discovering homicides or other deaths that involved questions of legal responsibility, an imperative to investigate deaths that intersected

---

[35] The census-taking procedures implemented by the Beijing police, as described by Sidney Gamble, match the provisions included in national regulations, promulgated in 1915, that governed police authorities' maintenance of records on the demographic composition of urban populations. Compare Gamble, *Peking*, 91–3 with the regulations in Cai Hongyuan, *Minguo fagui jicheng* [Collection of laws and regulations in Republican China] (Hefei: Huangshan shushe, 1999), 13.229–33. As Tong Lam has shown, registering at least some segments of Beijing's population was one of the earliest projects implemented under the new policing institutions established during the New Policies reforms. Lam, "Policing the imperial nation," 901–3. Also see Hou Yangfang, "Minguo shiqi quanguo renkou tongji shuzi de laiyuan" [On the sources of national population statistical figures during the Republican period], *Lishi yanjiu*, no. 4 (2000): 3–16.

[36] Dray-Novey, "Spatial Order and Police in Imperial Beijing," 897–903.

[37] As early as 1908, the hygiene department of the Ministry of Civil Affairs had collected statistics for purposes of epidemic prevention. These included information on sex, age, date of infection, and cause of death for those who died in the Inner and Outer Cities of Beijing. You Jinsheng, "Qingmo Beijing neiwaicheng jumin siyin fenxi" [An analysis of causes of death of residents of the Inner and Outer Cities in Beijing at the end of the Qing], *Zhonghua yishi zazhi* 24, no. 1 (1994): 23–4. For data on morbidity and mortality collected by the police as of the late 1910s, see Gamble, *Peking*, 115–8, 417–9. For subsequent attempts to collect vital statistics in Republican Beijing, see Cameron Dougall Campbell, "Chinese Mortality Transitions: The Case of Beijing, 1700–1990" (PhD Dissertation, University of Pennsylvania, 1995), 133–42.

with the other goals of surveillance through a police-run system of death-reporting and burial permits.

In order to implement these different forms of supervision over the deaths of city dwellers, the Beijing police enforced an arrangement through which the family of anyone who died in the city was required to report the death to the proper authorities in order to receive a pass that would allow one to take the body through the city gates. The gates had long been used as a point at which the officials who policed the imperial capital had controlled movement into and out of the Inner and Outer Cities and gathered information about individuals who passed through and the goods that they were carrying with them.[38] In a report on the sanitary conditions of mid-nineteenth-century Beijing, John Dudgeon (1837–1901) described a system of gate passes that is remarkably similar to that which the police in Republican Beijing used to establish surveillance over death in the city for purposes of policing and public hygiene.[39] According to Dudgeon, anyone wishing to take a body through the city gates had to be in possession of a ritual document called a *yangbang* (a word that means, literally, "notice of misfortune"), which was issued by the yinyang master whom the mourning family employed to manage the timeline and sequence of funeral arrangements.[40] The document itself outlined the dates on which encoffining and burial should be carried out, as well as various prohibitions that the family would have to respect in order to avoid further calamities. Such a document served the purpose of a "gate pass," in Dudgeon's words, when carrying the body out of the city. In cases for which a death involved homicide or suicide, the family could only obtain a gate pass following capital officials' investigation of the death.[41]

In 1920s and 1930s Beijing, there were two kinds of passes that one could obtain: the burial permit (*taimai zhizhao*) issued by police and the procuracy, or the funeral permit (*chubin zhizhao*) that the city's new hygiene authorities began to issue in the 1930s. In a city as heavily policed as Beijing, possessing such a permit not only allowed one to carry a body

---

[38] Dray-Novey, "Spatial Order and Police in Imperial Beijing," 895; Dray-Novey, "The Twilight of the Beijing Gendarmerie, 1900–1924," 359.

[39] *Medical Reports*, 31.

[40] For more on the writing of *yangbang* (also referred to as *yangshu*), see Li, *Beiping fengsu leizheng*, 131, 185, as well as an anonymously penned feature describing funeral customs in SJRB April 15, 1932, 8. The latter seems to have been one of the sources on which Li Jiarui based his own description. Also see Chang, *Lao Beijing de fengsu*, 207.

[41] *Medical reports*, 31. Also see a memorial submitted by Jiangxi Circuit Investigating Censor Liangbi in 1886, which addressed the malfeasance of capital officials in delaying the provision of passes to families after forensic examinations. Memorial of Liangbi, 6/5/ GX12 (1886), First Historical Archives of China (FHA) 3-110-5688-39.

through the city gates for burial, but would have been important as well
for demonstrating the legality of one's claims over a body to any inquiring
police officer.[42] In order to obtain such a permit, the family or acquaintances of a deceased person had to present one of several kinds of documentation to police of the district in which one lived. The immediate
purpose of doing so was to prove that there was nothing untoward or
suspicious about the circumstances surrounding the death and, by implication, that there was no need for police and judicial authorities to
investigate further. We can get a sense of the different forms of death
documentation that were acceptable to the police from a description of
the system given by city hygiene officials in the early 1930s:

When urban residents go to sub-stations of the PSB to report a death, according to
usual practice they must possess a death certification document, of which there
are three kinds. The certificates issued by yinyang masters are commonly used.
In the case of adherents of Islam and Christianity, staff from the respective place
of worship issues a certificate. As for those who die in a hospital, then the hospital
issues a death diagnosis certificate. Such long-standing measures have been
passed down to the present and are used to check whether there are deaths for
which the circumstances are unclear.[43]

It was common well into the 1930s for families applying for burial permits
to simply use the documentation provided by the yinyang master whom
they employed to coordinate the funeral arrangements. As Yang Nianqun
has shown, aside from handling this ritual work, for which they were paid
by the family, the yinyang master also certified the cause of death based on
observation of the social relations surrounding the deceased and drug
prescriptions that had been used before death, as well as an inspection of
the body.[44] Yinyang masters were prohibited from issuing death certification documents in cases involving "unnatural deaths (*biansi*) or in which
the cause of death is unclear" – that is, the very cases that were supposed
to fall under the jurisdiction of the city procuracy.[45] Prior to the late
1930s, it would have been much less common for a family to present

---

[42] Gamble, *Peking*, 115; *Beiping shi weishengju dier weishengqu shiwusuo nianbao* [Annual
report of the Second Health District Station of the Beiping Bureau of Hygiene] (Beiping:
Beiping shi weishengju dier weishengqu shiwusuo, 1935), 14.

[43] *Beiping shi zhengfu weishengchu yewu baogao* [Work report of the Beiping Municipal
Government Hygiene Office] (Beiping: Beiping shi zhengfu weishengju, 1934), 74–5.

[44] Yang Nianqun, *Zaizao "bingren": Zhongxiyi chongtuxia de kongjian zhengzhi, 1832–1985*
[Re-Making "Patients"] (Beijing: Zhongguo renmin daxue chubanshe, 2006), 160–4.

[45] Ibid., 160. As Yang notes, this rule was included in an early set of regulations governing
the work of yinyang masters, "Regulations of the Metropolitan Police Board on the
supervision of yinyang masters," 1913, BMA J181-18-222, 11–5. For cases in which
yinyang masters found suspicious signs on a body and notified police, see SJRB
January 11, 1934, 8; SJRB December 24, 1935, 6.

a medical death certificate as part of their application for a burial permit. For proponents of Western scientific medicine, the idea that deaths should be certified by physicians or hospital personnel had significance for clinical and public health work that greatly exceeded its practical utility for detecting cases of unnatural death and homicide. In the trial public health office that was created in the Second Inner Left district in 1925 through the cooperation of PUMC and the city police, the use of medical death certificates expanded as physicians and trained health officials affiliated with this office increasingly took over the work of investigating deaths in order to collect usable mortality data.[46] By the late 1920s, burial permits in this district were routinely being issued by the police on the basis of medical death certificates.

Over subsequent decades, statistics investigators (*tongji diaochayuan*) working under the auspices of the city government expanded the medical certification of deaths into other districts of the city, collecting mortality data according to a simplified cause-of-death classification scheme that had been deployed in the Inner City health district by John B. Grant (1890–1962) and his colleagues as a practical measure for Chinese police and hygiene authorities with rudimentary training to be able to determine cause of death in ways that were compatible with internationally recognized categories.[47] By the late 1930s, these personnel had supplanted the death certification work of yinyang masters, who lost the legal authority to carry out death certifications in early May 1936.[48] From this point onward, family members of those who died of natural causes had to report the death to district police authorities, who would notify hygiene officials to investigate the case. Through this arrangement, the city successfully implemented a system for collecting mortality data based on Western medical disease classifications that conformed with the Nationalist state's regulations on the collection of vital statistics, promulgated by the

[46] For the establishment and impact of the health station, see Bullock, *An American Transplant*, 144–51. For the ways in which the everyday meanings of birth and death were transformed under this station, as well as under the health offices established in other city districts, see Yang, *Zaizao "bingren,"* 110–18, 127–73. For the death certification procedures used in the health station, see monthly and annual reports collected in "China – Peking Health Demonstration Station, Reports, 1924–1926," Rockefeller Foundation records, International Health Board/Division Records, RG 5 (FA 115), Series 3, Box 218, Folder 2734. Rockefeller Archive Center (RAC).

[47] Campbell has noted that city hygiene authorities used this classification scheme into the 1940s. Campbell, "Chinese Mortality Transitions," 141–2.

[48] This occurred after years of debate and tentative pronouncements surrounding the replacement of yinyang masters' death certifications with a system of medical death registration. Yang, *Zaizao "bingren,"* 143–4, 164–6; SJRB March 5, 1931, 7; SJRB November 23, 1933, 8; SJRB December 2, 1933, 8; SJRB December 27, 1934, 8; SJRB January 31, 1935, 8; SB May 9, 1936, 4; SJRB May 9, 1936, 6; SJRB May 10, 1936, 6.

Ministry of Interior in October 1928, which mandated that physicians or hygiene officials carry out this work.[49] This shift in death investigation procedures had implications for the procuracy's forensic work, as we will see. From the mid-1930s onward, statistics investigators increasingly handled the investigation of bodies discovered in public spaces, cases in which someone died from illness in public, or other cases in which a death was unexpected but with little criminal suspicion. Prior to this point, such deaths had constituted a significant component of the procuracy's forensic caseload and one that had, in this office's own estimation, distracted its officials from their central mandate to investigate and prosecute homicide cases.

These procedures for the compulsory reporting of deaths represented a fairly limited state involvement in the management of death. It was not the ambition of city officials to fundamentally change urban death practices or the ideological content of the rituals themselves, but simply to make city dwellers' responses to death more compatible with the needs of the state. In this connection, we might compare the efforts of Beijing city authorities during the 1920s and 1930s with the much more thorough attempts of local and national authorities to transform funeral and burial customs that were carried out following the communist revolution in 1949.[50] These later efforts included the widespread implementation of cremation over burial and new ritual forms such as the memorial meeting (zhuidao hui), as well as a general repudiation of older funerary rituals, specialists, and meanings.[51] These later shifts reflected an expansion and institutionalization of earlier critiques of old funeral customs, already voiced during the Republican period, which focused on the wasteful expenditures involved in complex and lavish funeral rituals, the haphazard and uneconomical use of land that was the result of private burials, and concerns about the hygienic implications of delaying the disposal of the corpse.[52] The early PRC's nationalization of the funeral industry and other attempts to regulate funerals and the disposal of remains involved

---

[49] Zhang Zaitong and Xian Rijin, *Minguo yiyao weisheng fagui xuanbian, 1912–1948* [Selected laws and regulations on medicine and hygiene in Republican China, 1912–1948] (Jinan: Shandong daxue chubanshe, 1990), 170–1.

[50] During the late 1920s and 1930s, the Nationalist state likewise attempted to reform funeral customs, albeit with less widespread or consistent effect than the deeper reforms of the post-1949 period. Rebecca Nedostup, *Superstitious Regimes: Religion and the Politics of Chinese Modernity* (Cambridge: Published by the Harvard University Asia Center. Distributed by Harvard University Press, 2009), 251–8.

[51] Martin K. Whyte, "Death in the People's Republic of China," in *Death Ritual in Late Imperial and Modern China*, ed. Watson and Rawski; Liu, "Dying Socialist in Capitalist Shanghai," 52–97.

[52] For critical assessments of funeral customs in Shanghai, see discussion in *Libao* February 21, 1937, 6. Also see Nedostup, *Superstitious Regimes*, 251–3.

a significant and seemingly unprecedented expansion of state involvement in reorganizing the societal response to death.

There were certainly instances in which city authorities in Republican Beijing acted to reform or regulate funeral customs, and these tended to revolve around questions of hygiene. For example, there were periodic attempts to restrict the practice of delayed burial or to plan the building of new-style "public cemeteries" (*gongmu*) that would be more hygienic and economical than existing burial grounds.[53] Yet, broadly speaking, during the 1920s and 1930s it was not the ambition of the Beijing police or other municipal agencies to fundamentally change how city dwellers handled their dead. Even in the early 1930s, a point when the city's yinyang masters were being criticized for hindering the creation of a medical system of death certification, the head of hygiene in the city, Fang Yiji, still stated publicly that his office did not intend to prohibit city dwellers from employing ritual specialists in funerals given that this was a question of death customs that lay beyond the jurisdiction of this agency.[54] Indeed, in most cases police and hygiene officials simply worked around the existing set of rituals, building institutional procedures that required contact with the state (usually, police of the district in which one lived) without creating overwhelming burdens on a family or significantly disrupting the timing of a funeral or burial. By contrast, the nature and extent of state involvement was different for those deaths that had not been certified by a yinyang master or hygiene investigator, a broad category of death cases that included homicides, suicides, fatal accidents, and bodies discovered in public. For much of the Republican period, these deaths became the object of an important if unwieldy area of urban governance that fell within the professional jurisdiction of police and the procuracy.

Suspicious Deaths

Like other aspects of policing and the administration of justice in the city, the role of police and judicial authorities in the investigation of violent and

---

[53] For delayed burial, see SJRB June 16, 1931, 7. For a subsequent attempt by hygiene authorities to implement a long-term plan to outlaw the practice, see SJRB June 28, 1937, 5; SJRB July 5, 1937, 5. Beginning in late 1934, the Bureau of Social Affairs, along with other municipal agencies and self-government officials in the city, began to lay the groundwork for establishing city-maintained public cemeteries (*gongmu*) in the suburban districts surrounding Beiping. For communications, reports, and surveys from this process, see BMA J2-7-117. Also see SJRB February 27, 1936, 6. For the larger context of Nanjing-decade cemetery reform, see Nedostup, *Superstitious Regimes*, 254–8.

[54] SJRB December 3, 1933, 8.

unexpected deaths had late imperial precedents.[55] Under the complex arrangement of overlapping policing agencies that had overseen the Qing capital, officials who administered jurisdictions known as the Five Wards had primary responsibility for investigating cases of suspicious death and homicide.[56] These officials also examined the corpse in cases of those who were found dead on the streets and in other urban spaces from illness, starvation, or exposure, a condition often referred to as *daowo* – literally, "fell down flat." Regulations specified that officials of the Five Wards could resolve on their own authority non-homicide cases involving those who "fell down dead in the road" (*daotu daobi*) or "died from illness in inns" (*kedian bingwang*), after which they were expected to bury the body.[57] In more serious cases, these officials were to conduct the investigation under the supervision of higher authorities within the Wards as well as the Board of Punishments, the central government ministry responsible for the administration of justice.

This old area of urban policing was reorganized during the New Policies reforms as modern police and judicial offices were established in the city. By the last years of the Qing, officials of the city procuracy were overseeing corpse examinations in the same kinds of cases that officials of the Five Wards had previously handled, including homicide cases as well as those in which someone had simply died of natural causes on the street. During the Republican period, the scope of death cases investigated by the procuracy came to be associated with a new term, *biansi* (unnatural deaths), one of many words in Chinese history that have evoked death under less-than-ideal circumstances, and one that usually applied to the same kinds of cases in which it could be said that one had "died prematurely" (*si yu feiming*).[58] Which kinds of death cases were actually included in this category was largely decided by the police, who

---

[55] For more on the interplay between older policing and legal practices and new ones based on Western models in Beijing during this period, see Ng, *Legal Transplantation in Early Twentieth-Century China*.

[56] For regulations on the administration of forensic examinations in the capital, see *Qinding Da Qing huidian shili, juan* 851, 810.367–9, 375–7, *juan* 1031, 812.357–8, and *juan* 1037, 812.398–401.

[57] *Qinding Da Qing huidian shili, juan* 851, 810.369, *juan* 1037, 812.399–401.

[58] For the latter, see, for example, SJRB November 4, 1930, 7. J.C. Hepburn's *A Japanese and English Dictionary; with an English and Japanese Index* (Shanghai: American Presbyterian Mission Press, 1867, 100) defined the Japanese word *henshi* (*biansi*) as "n. A strange, unusual, or unnatural death." Jia Jingtao explains the word as "a Japanese legal term. An unusual death suspected to involve criminal circumstances." Jia, *Shijie fayixue yu fakexue shi*, 285. For other concepts that were used to describe "bad deaths" caused by violence or that occurred under other kinds of non-normative circumstances in late imperial and twentieth-century China, see Zhang Ning, "Corps et peine capitale dans la Chine impériale: Les dimensions judiciaires et rituelles sous les Ming," *T'oung Pao* 94 (2008): 275.

maintained responsibility for reporting deaths that were in need of investigation to the procuracy. The elaborate institutional infrastructure for enforcing city dwellers' reporting of deaths to the police, as well as police officers' routine surveillance of the urban spaces in which bodies were found, thus provide important contexts for understanding the nature and scope of the procuracy's forensic work as it was carried out in practice. While the procuracy did investigate obvious cases of homicide, it was also routinely called upon to examine the body in cases that presented no criminal suspicion.

In some cases, the procuracy was called to the scene because relatives had improperly reported a death or had failed to obtain the necessary certification document. For example, on the night of July 18, 1934 a person named Zhao Lianzhong reported the death of a young relative who was temporarily staying in Beiping to police in the Outer First district of the city.[59] Upon questioning Zhao, as well as the manager of the shop in which he worked, district police discovered that the deceased, named Zhao Xidong, had suffered from consumption and was passing through Beiping on the way to his native Zaoqiang county, Hebei. The police subsequently notified the procuracy, which dispatched a judicial policeman and an inspection clerk to examine Zhao's body. Reaching a verdict that there were no wounds on the body and that the death was due to illness, these officials issued a burial permit. Zhao's body was subsequently buried at a public cemetery located near Guangqu Gate in the northeastern quadrant of the Outer City and the police ordered Zhao Lianzhong to notify Zhao Xidong's parents so that they could come to Beiping and claim the body. The police report describing Zhao Lianzhong's initial exchange with police indicates that there was very little about the circumstances of the death that was overtly suspicious. Rather, the problem was that even though Zhao Lianzhong, a relative, had reported Xidong's death to the police, he did not provide the documentation of a yinyang master or hygiene investigator that could prove that the death had been due to natural causes.

In other cases, the procuracy became involved because a death had occurred under circumstances that made it impossible for a relative or acquaintance to report it to the police in the first place. These cases could involve unexpected fatal incidents occurring in the streets and other spaces of the city or cases in which a body was discovered by passers-by. For example, on the same page of the January 9, 1928 issue of *Morning Post* with which this chapter began, one might have read about the

---

[59] Outer First District to Beiping PSB, July 20, 1934, BMA J181-20-15724, 2–4.

discovery of a body in a well at Wang Family Vegetable Garden in the southwestern corner of the Outer City:

Yesterday at about 9 o'clock in the morning, a neighbor of the surname Li took her daughter to draw water at the well with a bucket. As soon as they let down the wicker basket with their bucket, they saw a floating object and suspected at first that it was straw. Looking carefully, however, they saw that it was a corpse, even though they could not tell whether it was male or female. Pulling up the basket they took their bucket and ran back. The proprietor of the plot Wang Zhencai caught a glimpse of them from a distance and suspected that there was a peculiar matter given that woman and daughter were running away frightened. Upon inquiring into the matter further, he too was extremely surprised by the corpse that had been discovered in the well.[60]

Wang immediately reported the body to the nearest police sub-station, which notified the district police office. This office in turn requested that the procuracy carry out an examination of the body. In another case, reported in *Morning Post* on the following day, a patrolman stationed near the Tianqiao market area observed a rickshaw puller transporting a lone female passenger who was described for readers as having "neck supple and head hanging, saliva flowing onto her chest, and with ashen complexion."[61] Finding the sight of this woman suspicious, the police officer stopped the puller, discovering that the woman had expired from illness and that the male accompanying her had fled. The exact same three signs of an in-plain-sight dead body were reported in another *Morning Post* account, this time describing the case of a train attendant of the Beijing–Tianjin railway who discovered the body of a middle-aged man, fatally stricken by illness, left seated on a train car after all of the other passengers had left.[62] Station police soon notified district police authorities, who then contacted the procuracy, the typical pattern in such cases.

Given that bodies discovered in public places were typically given over to the jurisdiction of the procuracy, it should not be surprising that its officials and inspection clerks routinely examined the bodies of those who died of starvation, exposure, and, more generally, the severe poverty and social dislocation that afflicted Beijing during this period.[63] During the 1920s and 1930s, foreign and Chinese sociologists attempted to

---

[60] CB January 9, 1928, 7.   [61] CB January 10, 1928, 7.   [62] CB January 13, 1928, 7.

[63] For more on poverty and urban life during this period, see Chen, *Guilty of Indigence*. For the economic decline of Beijing, see Xu, "Wicked Citizens and the Social Origins of China's Modern Authoritarian State," 82–100. Also see Hugh Shapiro's discussion of the uncertainties of urban life in Beijing and the compelling argument that the prevalence of the medical disorder spermatorrhea was connected to these social and economic conditions. Hugh Shapiro, "The Puzzle of Spermatorrhea in Republican China," *positions: east asia cultures critique* 6, no. 3 (1998): 578–80.

investigate the causes and extent of urban poverty in Beijing and else-where in China and to classify and quantitatively analyze the socioeco-nomic composition of urban society through the collection of data on household income, expenditures, and other metrics.[64] While the analy-tical concepts and specific findings of these studies often diverged, in their totality they confirmed the overwhelming extent to which Beijing's popu-lation was weighted toward the lower ends of the socioeconomic ladder, however it might be defined. *Morning Post* and other newspapers routinely reported on the discovered bodies of "beggars" (*qigai*) who died from illness, starvation, or exposure, or poor people who "fell down dead" (*daobi*) in the streets, one of many patterns of urban life that were made visible in the newspapers.[65]

That bodies discovered outside figured so prominently in the caseload of the procuracy was the indirect result of the significant commitment of the police to burying the city's unclaimed dead, usually following the photographing of the body and issuing of a public notice to facilitate later identification.[66] Under the broad surveillance that they maintained over the dead on the streets and in municipal institutions, Beijing police officials were constantly receiving reports of deaths that fell outside of the normal channels of reporting and thus had not been certified. Yet, the police demonstrated a consistent reluctance to issue burial permits in these cases, systematically passing them on to the procuracy in spite of the repercussions that this arrangement had for this office (discussed in Chapter 2). In the late 1920s and early 1930s, the Beiping PSB routinely spent over 200 yuan per month to encoffin and bury the unidentified bodies of "poor people who fell down dead" (*daobi pinmin*) that were discovered throughout the city, as well as a smaller figure associated with burying the unclaimed bodies of those who were executed.[67] By the start

---

[64] Dong, *Republican Beijing*, 214–28; Chen, *Guilty of Indigence*, 48–62, 87–90, 94–6.

[65] CB November 24, 1925, 6; CB December 10, 1925, 6; CB December 11, 1925, 6; CB December 29, 1925, 6; CB June 24, 1926, 6; SJRB January 8, 1928, 7. Compare with Christian Henriot's discussion of the more reserved coverage of abandoned bodies in the foreign-language Shanghai press. Christian Henriot, "'Invisible Deaths, Silent Deaths': 'Bodies without Masters' in Republican Shanghai," *Journal of Social History* 43, no. 2 (2009): 409–10.

[66] Such a notice was issued, for example, in the case of the deceased woman discovered in the well at Wang Family Vegetable Garden. CB January 9, 1928, 7. For cases in which photographs were taken, see CB January 13, 1928, 7. Sometimes a corpse would be identified later on, as happened in the case of a 40 *sui* man who came to Beijing for medical treatment but collapsed dead in the street, only to be identified hours after the investigation concluded. CB June 11, 1926, 6.

[67] In August 1930, for example, the former cost 248 yuan and the latter 34 yuan. See First Department of Beiping PSB to Third Department, November 3, 1930, BMA J181-20-4148, 5–7. For monthly expenditures on coffins in all city districts during the period 1929–32, see correspondence of the First Department of PSB headquarters, which

of the Republican period, there were already long-standing precedents for local officials and publicly -minded associations and individuals to handle the task of burying those who died during political and military disorder or natural disasters, whose bodies were discovered in public places and without claimants to handle burial, or whose families simply lacked the resources for proper funeral arrangements.[68] During the early twentieth century, these were joined by new organizations such as the Chinese Red Cross Society, which buried those who died during the overthrow of the Qing in 1911–12 and in the subsequent military and political upheavals of the early Republican period.[69]

The significant commitment of the police in Beijing to burying the dead of the city was not primarily informed by ritual imperatives of the kind that shaped the work of benevolent associations, which buried unclaimed bodies and provided coffins to the poor as part of a broader mission of promoting orthodox patterns of funeral ritual and burial.[70] The immediate goals of policing the dead in Beijing were secular ones, generated by the practical challenges of governing the city. Aside from having an interest in maintaining the hygienic conditions of city spaces, the police in Beijing also showed a significant concern that unclaimed bodies remain accountable to some degree of bureaucratic control and available should relatives come forward to claim the body later on. Given that the city did not have a centralized morgue to which unidentified bodies could be brought and stored for later identification, public burial

routinely attempted to verify the amounts of these expenditures: BMA J181-20-4095-4099, BMA J181-20-4146-4151, BMA J181-20-6611-6622, BMA J181-20-9016, BMA J181-20-9345, BMA J181-20-12138. The average monthly expenditure on coffins in these files, which cover 26 month-long periods, was approximately 254 yuan, with a minimum 101 yuan and maximum 438 yuan. The basic arrangement throughout this period seems to have been that police districts provided the coffins and other expenses up front, subsequently seeking reimbursement from police headquarters on a monthly basis. In mid-1933, it reportedly cost 4.60 yuan to encoffin, transport, and bury every person who died at the Beggars' Home (Qigai shourongsuo), a municipal institution operated by the city's Bureau of Social Affairs and a place to which the procuracy's forensic examiners were dispatched. SJRB June 9, 1933, 8. In early–mid 1937, it cost police authorities in the Inner Third district 5.40–7 yuan to dispose of the unnamed corpses of those found dead in the streets: 3–5 yuan for the coffin, 1–1.60 yuan to transport and bury the body and to erect a marker, and 0.80–1 yuan to photograph the corpse for later identification. See lists contained in BMA J183-2-10445.

[68] Liang Qizi, *Shishan yu jiaohua: Mingqing de cishan zuzhi* [Doing good and moral improvement: charitable organizations during the Ming and Qing] (Taipei: Lianjing chuban shiye gongsi, 1997), 217–38; Meyer-Fong, *What Remains*, 128–32; Henriot, "Invisible Deaths, Silent Deaths,".

[69] Caroline Reeves, "Grave Concerns: Bodies, Burial, and Identity in Early Republican China," in *Cities in Motion: Interior, Coast, and Diaspora in Transnational China*, ed. David Strand, Sherman Cochran, and Wen-hsin Yeh (Berkeley: Institute of East Asian Studies, University of California, 2007).

[70] Liang, *Shishan yu jiaohua*, 218–24.

grounds were the de facto site at which such bodies were kept and from which they might have to be retrieved.[71] The importance of maintaining the "city's appearance" (*shirong*) in what was the national capital until the late 1920s also undoubtedly played a role in police authorities' concern that bodies be collected from city streets and transported out of the walled city and into outlying cemeteries.[72]

Officials of the procuracy might also be dispatched to examine the bodies of those who died in the network of poor relief, detention, and drug detoxification facilities that were run by the city, a customary requirement that routinely put the city's forensic examiners into contact with the most desperate of Beijing's population. For example, one of the grim institutions in which the destitute of Beijing died was the Vagrants' Clinic (Liumin yangbingsuo), a municipal facility located in a narrow alley to the south of Qian Gate. This clinic, run on the basis of meager financial support provided by the Outer Fifth district, provided rudimentary medical care, as well as shelter and porridge, to ailing individuals who were collected from the streets or sent from other municipal institutions. Empty coffins were left in the open, ready and waiting for the clinic's patients, a reminder of the role that this facility played as a place for dying.[73] In early February 1935, a reporter from *World Daily* toured the facility and interviewed its administrator, describing in detail the two rooms in which patients received treatment. This was also the site at which a number of those who lived on the margins of society spent the last hours of their lives:

While the manager was saying this, we had already gone through the entrance to the clinic. This was comprised of two rooms with half of the space covered in

---

[71] For example, in the case of a rickshaw puller who was mortally injured in October 1924 after falling down while trying to pull a passenger through the mud, because it was found that he "had no relatives in the capital," police encoffined his body and buried it at South Ridge Charitable Cemetery (Nan'gang yidi), a burial ground located beyond Yongding Gate at the southernmost border of the Outer City; this was the same site at which police buried the unclaimed body of a woman who was killed when she touched exposed electrical lines two months later. CB October 14, 1924, 6; CB December 11, 1924, 6. This area contained at least three distinct burial grounds that were managed by the city PSB and were used for burials at public expense. By the mid-1930s, it was reported that 10,703 burials had already been made: BMA J2-7-117, 137. Regarding the retrieval of bodies, see a case in which police of the Southern Suburban district responded to the request of a Mrs. Wang née Ma who wanted to retrieve the body of her husband, who had been executed a month earlier and subsequently buried in one of the burial grounds outside of Yongding Gate. Southern Suburban district to Beiping PSB, September 8, 1935, BMA J181-20-22085, 6–7.

[72] Chen, *Guilty of Indigence*, 100.

[73] SJRB February 11, 1935, 8; SJRB October 8, 1936, 6; SJRB October 9, 1936, 6. This facility is listed a number of times as the site for body examinations in a list of trips that forensic examiner Yu Yuan took around the city in December 1927, discussed in Chapter 2. See BMA J181-20-655.

heated sleeping platforms with a layer of reed matting on top. The walls were covered over with old newspaper that had already turned a dusky yellow color, and mucus and other foul substances were smeared all over it. The corpse of a dead male was lying rigid on one of the sleeping platforms, stark-naked except for a pair of short trousers, and about to be put in one of those flimsy coffins with an unmaintained burial ground outside of Yongding Gate as this person's final resting place. How pitiful! How sad![74]

That the Beijing procuracy routinely investigated deaths such as the one depicted in this article gives a sense of the broad scope of death cases that fell within its jurisdiction. In the day-to-day work of this office, forensic investigation was less about solving difficult cases – mysteries of the kind that one could read about in detective novels (*zhentan xiaoshuo*), for example – than in certifying the cause of death in a never-ending stream of cases, many of which involved death from natural causes or fatal accident. While it is clear that a broad swath of death cases was given over to the procuracy, it is surprisingly difficult to quantify what proportion of deaths in the city was treated in this way. Police, hygiene officials, and the procuracy had different goals in investigating deaths, used different cause-of-death classifications, and relied on different systems for maintaining records of the cases that they handled.[75] Moreover, when examining the considerable body of mortality statistics produced by police and hygiene officials it is usually unclear whether the numbers already accounted for deaths handled by the procuracy and, if they did, how consistently this was the case over time.

Nonetheless, one can compare different bodies of data collected during this period to get a sense of the proportion of urban deaths – homicides, fatal accidents, and deaths from natural causes – that were subject to these body examinations. One might compare, for example, mortality statistics collected by the police with judicial statistics compiled by the Ministry of Justice. In the period 1914–17, police statistics indicate that, on average, 17,925 people died in the city annually. Statistics on the caseload of the Beijing procuracy and its two branch offices indicate that for the exact same period an average of 3,759 cases of "killing and wounding" (*sha-shang*) were handled per year. If we use the latter figure as a rough and

---

[74] SJRB February 11, 1935, 8.

[75] For example, when in late 1947 the city Bureau of Hygiene requested that the Beiping procuracy complete monthly tabulated reports of the unnatural death cases that it handled in order to fill lacunae in hygiene authorities' mortality statistics, the procuracy demurred, noting that it only ever completed the examination forms required by the judiciary, had never compiled the data into tables, and that if the Bureau of Hygiene needed such information, they were welcome to "send personnel to our office to look through the files and copy the records." Beiping Bureau of Hygiene to Beiping Local Court, November 6, 1947, BMA J5-1-1278, 175-7.

imperfect proxy for the forensic cases handled by this office, we can calculate that it investigated about 20 percent of deaths at the absolute maximum, a figure that is undoubtedly high given that the judicial statistics on which it is based included non-fatal wounding cases in addition to homicides.[76] At the same time, this figure seems not to include a substantial component of the procuracy's forensic caseload: unexpected deaths or discovered bodies that were not homicides. During the 1920s and early 1930s it was not unusual for police throughout Beijing to encounter 100–200 cases involving those who "fell down dead" (*daobi*) in public places per month, and it is likely that most of these were reported to the procuracy for investigation.[77]

Bad Deaths as Public Spectacle

The body examinations carried out in these cases were public events that could attract crowds of people. That officials of the procuracy often did not arrive until hours after a death left time for passers-by and neighbors to form a crowd at the scene of an incident. In the case of a rickshaw puller who died in October 1924 after injuring himself while transporting a passenger, for example, it was reported that pedestrians "stood on the road and gathered around to watch, carts and horses stopped up the road, and traffic was blocked."[78] In the killing of Su Baojie, a particularly devious case of automobile hit-and-run discussed in Chapter 4, it was reported that between 1,000–2,000 people were present at the border between Taiping Lake and the campus of Republican University to observe the second examination of Su's remains.[79] And it was precisely such a crowd that drew the attention of a *Morning Post* reporter who went on to write the account of an attempted suicide that occurred

[76] For these figures, see Gamble, *Peking*, 417, and Sifa bu zongwuting diwuke, *Di yi ci xingshi tongji nianbao* [First annual report of criminal statistics] (n.p.: Gonghe yinshuaju, 1917), 17–24; Sifa bu zongwuting diwuke, *Di er ci xingshi tongji nianbao* [Second annual report of criminal statistics] (n.p.: Gonghe yinshuaju, 1918), 17–24; Sifa bu zongwuting diwuke, *Di san ci xingshi tongji nianbao* [Third annual report of criminal statistics] (n.p.: Gonghe yinshuaju, 1919), 17–24; Sifa bu zongwuting diwuke, *Di si ci xingshi tongji nianbao* [Fourth annual report of criminal statistics] (n.p.: Gonghe yinshuaju, 1921), 17–24.

[77] City-level data on cases involving those who "fell down dead" (*daobi*) in public spaces were collected unevenly. For several lists of monthly figures, see "List of the numbers of poor people who fell down dead as reported by each district in February 1921," BMA J181-18-13020, 9–25, which lists 170 deaths for the month; "Statistical table of those who fell down dead in each district for February 1930," BMA J181-16-208, 1–2, which lists 213 cases involving those who died of natural causes; "Statistical table of those who fell down dead in each district for April 1930," BMA J181-16-208, 5–6, which lists 109 such cases.

[78] CB October 14, 1924, 6.   [79] CB March 9, 1928, 7.

in November 1924, one of many instances in which the work of police and procurators was featured in the new realm of newspaper-driven commentary on city life that was disseminating images of urban death to a burgeoning – albeit not quite "mass" – readership.[80]

By the early 1920s, for example, brief accounts of violent and unexpected deaths appeared daily in *Morning Post*, interspersed with other stories about city life, society, education, culture, and crime. This well-known newspaper and its literary supplement disseminated international, national, and local news as well as the intellectual and literary works of a number of May Fourth-era writers, including Lu Xun (1881–1936).[81] The forerunner of *Morning Post*, a newspaper named *Morning Bell* (Chenzhong bao), began publication in August 1916 under the auspices of Liang Qichao (1873–1929) and the politician and official Tang Hualong (1874–1918), another late imperial constitutionalist who would become a major political figure in early Republican politics under Yuan Shikai and later Duan Qirui.[82] Over the course of the 1920s, *Morning Post* was at the forefront of developments that transformed the practice of modern Chinese journalism. These included the move to publish literary supplements, the professionalization of journalism as an academic field, and an expanded notion of journalists' role in engaging the intellectual, social, and commercial interests of an unprecedentedly broad reading public.[83]

Featured in the pages of *Morning Post* were homicides and suicides; accidental deaths caused by automobiles, streetcars, electrical lines, and

---

[80] CB November 6, 1924, 6. Xiaoqun Xu estimates that those who subscribed to or otherwise purchased newspapers and periodicals in early 1930s Shanghai accounted for almost 16 percent of the population, a figure that would have been highly disproportionate to much of China, albeit less so for Beijing. Xu, *Chinese Professionals and the Republican State*, 47–8. Also see Leo Ou-fan Lee and Andrew J. Nathan, "The Beginnings of Mass Culture: Journalism and Fiction in the Late Ch'ing and Beyond," in *Popular Culture in Late Imperial China*, ed. David Johnson, Andrew J. Nathan, and Evelyn S. Rawski (Berkeley: University of California Press, 1985), 368–78. For a sense of some Beijing church members' newspaper subscription habits, see Gamble, *Peking*, 361, 508. Also see Gamble's discussion (pp. 147–56) of some of the public lecture halls and libraries that existed as of the end of the 1910s to provide Beijing city dwellers with opportunities to read newspapers without purchasing them.

[81] For an overview of the newspaper as well as discussion of the ideological orientation of its literary supplement, see Zhonggong zhongyang Makesi Engesi Liening Sidalin zhuzuo bianyiju yanjiushi, *Wusi shiqi qikan jieshao, di yi ji* [An introduction to periodicals of the May Fourth period, volume 1] (Beijing: Renmin chubanshe, 1958), v. 1, 98–143.

[82] The newspaper took the name *Morning Post* in December 1918 after *Morning Bell* was shut down earlier in the year following its coverage of Duan Qirui's acceptance of Japanese loans. The newspaper maintained this name until June 1928 when it ceased publication.

[83] For more on these shifts, see Timothy B. Weston, "Minding the Newspaper Business: The Theory and Practice of Journalism in 1920s China," *Twentieth-Century China* 31, no. 2 (2006): 4–31.

trains; and deaths from exposure and starvation – visible manifestations of the social displacement that defined urban life during this period of war and uncertainty. A basic element of this reporting was sensationalism, defined by historian Joy Wiltenburg as "the purveyance of emotionally charged content, mainly focused on violent crime, to a broad public."[84] Titles that previewed scenes of death and destruction were common: "Yesterday a streetcar ran over and killed a child" (subtitle: "A body run over into three pieces, too awful to look at!"), or "Beijing-Hankou railway train crushes a person's head."[85] Vivid images of fatal violence were common as well, as was the explicit description of scenes that were ostensibly transgressive of societal norms surrounding the handling and disposal of the dead body. Streetcars, trains, and other mechanized dangers of modern urban life dismembered and disemboweled grotesquely, creating scenes that were "too awful to look at" (*can bu ren kan*).[86] At the same time, of course, the journalists of *Morning Post* and other newspapers were invested in transforming these tragedies into objects of consumption and interest, if not entertainment. That such reporting often included a gruesome description of the corpses might also imply readers' interest in reading about dead bodies, however grotesque their representation on the page.[87]

One such object of reporting was the ever-prevalent "female corpse" (*nüshi*) that appeared at the center of cases involving murder, suicide, and unexpected death in *Morning Post* and other newspapers. For example, the account of a woman who died of natural causes at Runshen Female Bathhouse in early February 1927 draws the reader into a space – both corporeal and architectural – forbidden to men.[88] Once judicial authorities arrived at the scene, a female forensic examiner named Ms. Sun, one of the so-called "midwives" (*wenpo*) who assisted the procuracy in these investigations, conducted a preliminary inspection of the body, which was shrouded in a blanket. As *Morning Post* reported, "the midwife opened the blanket and saw many small wound-like marks on the chest and arms of the deceased," after which "the body was covered again" and questioning

---

[84] Joy Wiltenburg, "True Crime: The Origins of Modern Sensationalism," *American Historical Review* 109, no. 5 (2004): 1377.

[85] CB March 3, 1927, 6; CB January 11, 1928, 7.   [86] CB April 30, 1926, 6.

[87] In this connection, see Tie Xiao's discussion of necrophilic desire in *Cadaver* (Jianghai, 1927), one of the early writings of left-wing writer Hu Yepin. Hu's short story revolves around the desire that an anatomist discovers that he feels toward a corpse in the dissection room. Xiao's analysis is suggestive not only of the ways in which the corpse could be construed as an object of desire in the literary imagination, but also of one way in which explicit description of dead bodies could be deployed in literary texts. Tie Xiao, *Revolutionary Waves: Imagining Crowds in Modern China, 1900–1950* (Book manuscript), chapter 4.

[88] CB February 8, 1927, 6.

of bathhouse employees proceeded. The deceased woman had been discovered collapsed on the floor in one of the bath areas. When a physician was summoned from a nearby hospital, the body had to be moved to an outer area because "men were not allowed in the bathing area." In the end, this investigation yielded that the woman had simply died from illness and that the marks were the result of the fact that the skin had recently been scraped for therapeutic purposes. When reporting on the investigation of women or girls who died, it was not uncommon for *Morning Post* to describe in vivid detail the forensic examination of the "lower body" (*xiati*) of the deceased, the very area of the body that official procedure maintained should be shielded from the view of onlookers and even the investigating officials.[89] In a case from early May 1924 involving the murderous beating of a young woman who had been kidnapped and forced into prostitution, even the fact that the victim's bound feet were bare when the body was disinterred was not left to imagination.[90]

Perusing the pages of *Morning Post* and other Beijing newspapers, one finds that the "tragedies" most commonly described were of corpses destroyed and dismembered, not abandoned. It was rare for the focus of reporting to be the neglect of corpses. There were exceptions to the efficient and orderly disposal of dead bodies as it was portrayed in *Morning Post*, but they were uncommon. In the early morning hours of March 8, 1928, a military officer's driver struck a rickshaw in the warren of narrow lanes lying to the southwest of Qian Gate in the Outer City.[91] In speeding away from the scene with a brave patrolman hanging on the side of the automobile yelling at the driver to stop, the driver plowed the car into the rocky surface of a construction site. This caused an attendant who was riding in the car to be catapulted onto the rocks, where he received fatal injuries. The family of the dead attendant refused to encoffin the body following the forensic examination – a sign of protest against those whom they felt were responsible for the death – with the result that the corpse "lay exposed on the street all night once again."[92] It was subsequently reported in *Morning Post* that the owner of the car and the driver who caused the death provided compensation as well as a coffin. Only then did the relatives of the deceased allow the body to be

---

[89] CB July 16, 1924, 7.

[90] CB May 7, 1924, 6. For more on the complex meanings of the exposed bound foot as a body part that was viewed as "taboo" in late imperial times but that had become increasingly visible in anti-footbinding texts and public displays by the early twentieth century, see Dorothy Ko, *Cinderella's Sisters: A Revisionist History of Footbinding* (Berkeley: University of California Press, 2005), 41–3.

[91] CB March 9, 1928, 7.   [92] CB March 10, 1928, 7.

encoffined. It was brought to Five Sages Temple (Wusheng an) in the southwestern quadrant of the Outer City to await later burial.

Such cases were extraordinary, however. The overwhelming image of death in the city that was portrayed in *Morning Post* was that of an efficient state response to urban fatality. Long-time readers would have expected that the discovery of a body or a fatal incident would be followed by the arrival of the police. Personnel from the procuracy would arrive next for the examination of the body, followed by the issuing of the burial permit. Finally, the body would be claimed by relatives or others or buried by the police themselves. The journalistic conventions of these accounts, which very clearly required the reporting of each of these steps, would have given readers the not inaccurate impression that these procedures were followed in every case regardless of the social status of the person whose dead body happened to fall under the gaze of police and judicial officials.[93]

This was a period in which Chinese state policies toward the poor changed considerably in Beijing and elsewhere. Janet Chen has charted the rise of an "extractive notion of social citizenship" in which the urban poor came to be viewed as a potential source of labor for the national economy that could be exploited through detention in workhouses and other quasi-penal institutions.[94] It is striking, in this context, just how much the interactions that city police and judicial officials had with those who lived on the margins of society *after* they died were not characterized by extraction or punitive insensitivity.[95] Rather, city officials generally disposed of the bodies of the poorest and most marginalized members of urban society in ways that were congruent with widely held expectations about how dead bodies should be treated while doing what they could to keep open the possibility that a relative could claim the body later on.[96] In attending to this area of urban administration, Beijing officials expended large sums of money as well as the time and energy of the police, procurators, and forensic examiners who were mobilized to handle these cases. The result, visible on the streets and in the newspapers, was

[93] For example, see CB November 24, 1925, 6, which mentions explicitly that the unclaimed corpses of two "beggars" (*qigai*) who froze to death were encoffined and buried with measures taken for later identification of the bodies.

[94] Chen, *Guilty of Indigence*, 10.

[95] Compare with Thomas Laqueur's depiction of the "pauper funeral," the intentionally minimalist funerary arrangements provided for those who died in nineteenth-century England's regime of workhouses. These demonstrated, Laqueur argues, official authorities' endorsement of a form of burial signifying indignity and "social worthlessness." Thomas Laqueur, "Bodies, Death, and Pauper Funerals," *Representations* no. 1 (1983): 109–31.

[96] On this point, see discussion of the police procedures surrounding medical schools' claiming of bodies for dissection, discussed in Chapter 6.

an area of intensive state society interaction in which a particular set of officially endorsed death practices was widely enforced throughout the city.

Conclusion

The sensational image of accidental death with which this chapter began contrasts sharply with the mode of representing death that one finds in texts that focus on customs (*fengsu*) and ritual as the main sites at which the living encountered the dead in Beijing. We find this perspective, for example, in folklorist Li Jiarui's *Beiping fengsu leizheng* (A categorized collection of Beiping customs, 1937), a work that described various aspects of social and cultural life in Beijing through brief anecdotes mined from a wide range of historical and contemporary sources. The work included a chapter on "weddings and funerals" (*hunsang*) with largely idealized descriptions of various ritual acts that might be carried out after death.[97] In fact, as we have seen, this more conventional world of death customs and the images of tragic death and homicide investigation that were disseminated daily in the newspapers were not as distant as they might appear at first glance. Beijing city dwellers carried out funerals within the context of police authorities' varied forms of surveillance over the dead. Under this arrangement, reporting a death to the authorities was an administrative requirement that had to happen before the ritual acts ending in burial or the storage of a coffin could be completed. One could even say that the Beijing police had co-opted particular elements of these rituals for the purposes of extending its own reach over urban society.

It is in this context that we might revisit the question of just how standardized Chinese death practices were, less for what this can tell us about earlier debates over the role of ritual praxis in Chinese "cultural integration"[98] than for what it can reveal about the ways in which new "standardizing forces of modernity" – a modern state, new professions, and new commercial forces – transformed the meanings of death in China, much as they did in other societies during this period.[99] The urban state's multifaceted interest in the deaths of city dwellers encouraged consistency if not uniformity in certain aspects of society's response to fatality. The procedures surrounding death-reporting and

---

[97] Li, *Beiping fengsu leizheng*, 128–32.

[98] James L. Watson, "The Structure of Chinese Funerary Rites: Elementary Forms, Ritual Sequence, and the Primacy of Performance," in *Death Ritual in Late Imperial and Modern China*, ed. Watson and Rawski.

[99] Bernstein, *Modern Passings*, 10.

burial permits blurred the boundaries between administrative procedures and ritual practices, creating an area of governance in which city dwellers routinely cooperated with police in the administration of the city while carrying out ritual acts that carried normative cultural weight. In these ways, the urban state did play a role in shaping normative expectations about what should happen after death and how bodies should be handled. Yet, it did so not through direct interference in funeral rituals, but rather by requiring easily accomplished practices of death-reporting as well as by regulating, in a very rudimentary way, the transit of bodies in urban space. Indeed, it is striking how commonplace city dwellers' reporting of deaths was during this period, a point that suggests that this was one area in which the police were quite successful at disciplining everyday behaviors within urban society.

In newspaper reporting on urban fatality, we also find congruence if not complicity with these practices of urban policing and everyday discipline. Newspaper coverage implicitly argued for the legitimacy of the state's involvement in the aftermath of death by portraying the policing of the dead as a routine, normal, and expected part of everyday life. Moreover, the fatal incidents that became the subject of newspaper accounts were located in this reporting relative to police administrative districts, thus inscribing the administrative order of the police onto city neighborhoods and streets. While the circumstances that led to these deaths might change, as could the bizarre details, readers would have reasonably expected to see the same sequence of administrative procedures following each death that was reported. News reporting could support the state's policing of the dead in even more direct ways as well. In one case from late November 1929, a university student learned about his father's fatal collapse in the East Station on the previous day after reading about it in the newspaper, and subsequently went to identify and claim the body.[100] Likewise, when police of the Outer Fourth district discovered a hempen sack containing human bones, ostensibly lost by relatives who were transporting the body to burial grounds outside of the city, *World Daily* published the official police announcement asking the relatives to step forward and claim the body without fear of punishment.[101]

In all of these ways, death became more accessible to the urban state through its engagements with a wide range of ritual, commercial, and, more broadly, cultural practices. For cases involving violent or unexpected death, another source of urban officials' control and authority over the dead was the set of technical practices that the procuracy used

---

[100] SJRB November 21, 1929, 6; SJRB November 22, 1929, 6.
[101] SJRB November 12, 1934, 8.

to examine the body and establish cause of death. At the heart of the urban state's engagements with the body in such cases was a particular way of analyzing the wounds and other physical marks on the body and documenting the findings, a process that made the corpse accessible to police and legal authorities in distinctive ways. These were connected to a larger bureaucratic infrastructure, instituted during the Qing and maintained to a surprising extent during the Republican period, for systematizing the collection and interpretation of forensic body evidence. In Beijing and elsewhere, the legacy of the highly centralized forensic examination system of the Qing state thus remained a powerful force for standardizing the engagements between living and dead. Understanding what these techniques were, where they came from, and how they came to serve a state and society undergoing modern transformation requires that we turn to the procuracy and to the important work of the forensic examiners who inspected bodies on city streets, at crime scenes, and in municipal facilities and clinics.

2   On the Case with the Beijing Procuracy

On March 18, 1926, a large crowd of protestors gathered in front of the government offices of Duan Qirui, long-time power broker of the fractious militarist politics that had engulfed China for a decade and current Chief Executive of the Beijing government. Two days earlier, a coalition of Western countries and Japan had demanded that Chinese forces affiliated with the National Army of militarist Feng Yuxiang remove mines and other control measures that they had established around the mouth of the Hai River near Tianjin. This strategic area had been the site of recent altercations between Feng's forces and those of Zhang Zuolin (1872–1928), who had received the support of Japanese naval forces. In making this demand, the powers invoked provisions of the deeply unpopular Boxer Protocol that guaranteed them free passage from the sea to Beijing. Over the next two days, this demand aroused a tremendous amount of popular opposition that was quickly redirected onto Duan Qirui's government for its acquiescence. When a group of protestors arrived at the State Council on the afternoon of March 18, Duan's guards unexpectedly opened fire. In total, at least forty-three people were killed, many of them students, with about fifty, if not many more, receiving non-fatal wounds.[1] In the days and weeks following the massacre, students of Beijing's many universities took the lead in organizing memorial meetings and protests meant to mourn those who died as patriotic martyrs (*lieshi*) and to rally society to condemn "the country-selling government" (*maiguo zhengfu*) that had just traitorously massacred its own people.[2]

---

[1] Officials of the Beijing procuracy, which investigated the killings, personally examined the bodies of 43 individuals who were killed. CB April 3, 1926, 6. Regarding the numbers of wounded, Lin Benyuan, who documented the massacre and its aftermath, provided specific information about 55 wounded individuals who were treated in area hospitals. Lin Benyuan, *Sanyiba can'an shimo ji* [A record of the March Eighteenth massacre from beginning to end] (1926), appended to Fei Jingzhong, *Duan Qirui*. Jindai Zhongguo shiliao congkan di 90 ji (Taipei: Wenhai chubanshe, n.d.), 14, 16–7. The procuracy had itself examined the bodies of 45 wounded individuals. Lin also suggested, however, that "over 200 people" were wounded in the incident in the end. For more on the March Eighteenth massacre, see *Sanyiba yundong ziliao* [Materials on the March Eighteenth Movement] (Beijing: Renmin chubanshe, 1984).

[2] CB March 21, 1926, 6; CB March 25, 1926, 6.

Less than two weeks after the massacre, the Beijing procuracy released the results of its investigation into the killings, which were published in the Beijing press.[3] On the afternoon of March 18, an entourage from the procuracy had arrived at the scene of the killings. These officials oversaw examinations of the twenty-six bodies that police had collected from around the area and subsequently interviewed police and others who had been present at the scene.[4] The procuracy's investigation soon received the ambivalent sanction of Duan's government, which called upon the ministries of Army and Justice to investigate whether some of the guards had overreacted in discharging their duties. It also conformed to the wish of student protestors, who directly appealed to this office to pursue an indictment against Duan's cabinet for ordering the killings.[5] In the end, the findings from the procuracy's investigation directly contradicted the position of Duan's government that the guards had opened fire in self-defense while facing a violent crowd of communist-led agitators who were wielding pistols and other weapons.[6] As the procuracy explained in detail, this claim was not substantiated by the statements of witnesses, many of whom were members of the city police. One of the key findings of the investigation was that many of those killed had been shot in the back. As the newspaper *Morning Post* noted in its preamble to the procuracy's report, this finding demolished the notion that the guards had opened fire in self-defense.[7] In the end, because the actions of the guards purportedly fell under the jurisdiction of military justice, the procuracy gave the case over to the Ministry of the Army. Within weeks of the conclusion of this investigation, however, Duan Qirui was removed from office, first by Feng Yuxiang, whose own power was rapidly waning, and subsequently by Zhang Zuolin, who soon gained control of the capital.

Beijing's procurators and their forensic personnel played an important role in the investigation of cases involving death from violence or in which a city dweller had died under unexpected or suspicious circumstances. It routinely fell to these officials to judge how someone had died and who, if anyone, was legally responsible. In the process, they provided the evidence on which court judgments were made and impacted the lives of city people who became implicated in homicide investigations. Much as in the case of the March Eighteenth massacre, the broad legal and social impact of the procuracy in such cases rested on particular techniques for

[3] CB April 3, 1926, 6.
[4] CB April 3, 1926, 6; "Youguan sanyiba can'an sishang qingkuang de dang'an," 20–1.
[5] *Sanyiba yundong ziliao*, 82, 86, 102–3.
[6] For the government's statement in the wake of the killings, see *Sanyiba yundong ziliao*, 77.
[7] CB April 3, 1926, 6.

examining the dead body and establishing cause of death. These had developed within a larger infrastructure of institutions and procedures that was put into place during the Qing in order to standardize the use of forensic body evidence in criminal cases. Officials in early twentieth-century Beijing and elsewhere in China benefited from this system of practical knowledge, which could be applied systematically throughout the judiciary and which made the dead body legible to the modern state in consistent ways.[8] While the modernity of the *Washing Away of Wrongs* might be questioned by its medical reformer critics, as we will see, for city officials in Beijing it was part of a system of forensic knowledge that facilitated modern policing while supporting the goals of "formalization, standardization, and bureaucratization" within the judiciary itself, essential aspects of the project of modernization undertaken by proponents of legal reform since the last decade of Qing rule.[9]

Forensic Knowledge in Late Imperial China

Procurators in Beijing had a very direct relationship with the dead body, as we have seen. These officials and their subordinate body examiners routinely traveled to the site of a death and issued the burial permit that the family needed to transport a body out of the city for burial. That these officials personally oversaw such examinations without the assistance of physicians represented a significant point of departure from the foreign models on which Qing legal reformers had drawn. Procurators in Russia and Japan, for example, also had a great deal of authority over the collection and interpretation of physical evidence, but this hardly obviated the use of physicians in cases requiring forensic examination of a body. Indeed, both of these countries saw the development of robust communities of experts in modern Western forensic medicine who routinely asserted their authority in the law.[10] In Republican China, by contrast, it was less common for physicians to take part in this stage of criminal investigation, a reflection of the fact that there were relatively few

[8] My use of "legible" is inspired by James C. Scott, *Seeing Like a State: How Certain Schemes to Improve the Human Condition Have Failed* (New Haven: Yale University Press, 1998). Also see Tong Lam's discussion of the applicability of Scott's notion of "legibility" to the late imperial Chinese state and his compelling argument that "the transition from the so-called 'premodern' to the 'modern' should not be portrayed as the emergence of a legible and governable world. Rather, it should be understood as the shift from one form of legibility and governmental rationality to another as a result of a change in the political order." Lam, *A Passion for Facts*, 73.

[9] For more on this conception of "judicial modernity," as Xiaoqun Xu has defined it, see Xu, *Trial of Modernity*, 5.

[10] Becker, *Medicine, Law, and the State in Imperial Russia*, 193–209; Jia, *Shijie fayixue yu fakexue shi*, 292–4, 296–305.

physicians of Western medicine who were trained and willing to conduct these examinations. The limited involvement of physicians in forensic examinations also reflected the official policies of the judiciary, which actively maintained the authority of legal officials over this area of judicial practice. This arrangement was itself based on the practices of the legal system of the recently fallen Qing, which had invested the immediate authority over forensic inquiries with the local officials who were tasked with administering justice. In cases involving homicide, it was the county magistrate who was responsible for personally overseeing an examination of the body and ensuring that the right procedures had been followed.[11]

While the Qing state had not invested late imperial physicians with primary authority over forensic examinations, they did become involved in forensic cases in different ways. Fabien Simonis has found that physicians performed various tasks for county officials, including the treatment of prisoners, and routinely were called upon to confirm whether those involved in legal cases had suffered from madness.[12] Moreover, in a statute maintained from Ming law, the Qing legal code specified that in cases for which a quack (*yongyi*) was suspected of administering fatal treatment, officials could charge a physician with examining the drugs used and acupuncture points of the deceased.[13] Cases that appear in the late Qing legal and forensic literature indicate that there were instances in which officials made use of physicians' testimony, even though it is unclear how frequently this took place.[14] While physicians did become involved in some areas of forensic practice, the collection and interpretation of body evidence remained decisively under the purview of the

---

[11] For regulations that guided the administration of forensic examinations under the Qing, see *Qinding Da Qing huidian shili, juan* 851, 810.365–77. For an overview of the forensic procedures of the Qing and earlier dynasties, see Jia, *Zhongguo gudai fayixue shi*. For the general organization and operation of local government institutions during the Qing, see T'ung-tsu Ch'ü, *Local Government in China under the Ch'ing* (Cambridge: Council on East Asian Studies, Harvard University; distributed by Harvard University Press, 1988).

[12] Fabien Simonis, "Mad Acts, Mad Speech, and Mad People in Late Imperial Chinese Law and Medicine" (PhD dissertation, Princeton University, 2010), 493–8.

[13] Xue Yunsheng, *Duli cunyi chongkanben* [A Typeset Edition of the Tu-Li Ts'un-I with a Biography of the Compiler and Numbering and Titles Added to the Sub-Statutes] (Edited by Huang Tsing-chia. Taipei: Chengwen chubanshe, 1970 [1905]), 869; 297/00.

[14] For example, in an early Qianlong case that appeared in the late nineteenth-century *Collection of Rejected Cases*, higher authorities relied on the deposition of a physician, alongside other evidence, to argue that the biting off of a finger caused a wound that led to the victim's death. *Boan huibian* [Collection of rejected cases] (Beijing: Falü chubanshe, 2009 [1883]), 497. In an early Jiaqing case included in *Collected Evidence for Inquests* (Jianyan jizheng) and appended to expanded editions of the *Washing Away of Wrongs*, a physician was questioned regarding the odd condition of the wounds, which included a white scab not described in the *Washing Away of Wrongs*. See discussion in Will, "Developing Forensic Knowledge through Cases in the Qing Dynasty," 83, 88.

bureaucracy's officials – elite men who had passed the civil service examinations and had been appointed to positions in the empire's territorial
administration. In overseeing these examinations, these officials relied on
a standardized body of knowledge that was meant to routinize the handling of forensic examinations and facilitate provincial and central government officials' supervision over the forensic inquiries that were carried out
in localities.

One of the most important requirements was that conclusions about
forensic evidence accord with guidelines contained in the *Records on the
Washing Away of Wrongs, edited by the Codification Office* (Lüliguan jiaozheng Xiyuan lu), an official handbook that was completed early in the reign
of the Qianlong emperor (r. 1736–96) on the basis of earlier forensic
handbooks as well as a variety of other texts on medicine and related
subjects.[15] This official Qing version of the *Washing Away of Wrongs* was
similar in purpose and organization to the mid-thirteenth-century *Collected
Writings on the Washing Away of Wrongs* that was authored by the official
Song Ci, the source for substantial portions of the later text. This earlier
text is often characterized as the world's first "systematic" treatise on legal
medicine, given the broad scope of the topics covered in its pages as well as
the general commensurability between these topics and those of modern
legal medicine, including specialized subfields such as forensic entomology
and research on bodily decomposition.[16] The *Collected Writings* and its later
editions were divided into chapters addressing particular areas of forensic
practice (for example, "Examining bones" or "Washing and covering"
[*xiyan*], the name for a technique that was used to make bruises or wounds
visible on discolored body surfaces) as well as particular scenarios that
officials might encounter ("Suicide by hanging," "Beating or strangulation
made to look like suicide by hanging," "Death from drowning," "Death
from lightning strike," "Death from tiger attack").[17] While some of the

---

[15] For a compelling argument that the official *Washing Away of Wrongs* was completed in the
early 1740s, not 1694 as is usually claimed, see Chen, "'Xiyuan lu' zai Qingdai de
liuchuan, yuedu yu yingyong," 48–54.

[16] Jia, *Zhongguo gudai fayixue shi*, 65.

[17] It is worth quoting Pierre-Étienne Will at length on the organization of the official Qing
Washing Away of Wrongs: "It is a fact that in some of its parts *The Washing Away of Wrongs*
seems uselessly complicated, one reason for this complication being, as I see it, that the
late seventeenth-century editors of the official version, who set the pattern for the rest of
the dynasty, were hopelessly caught between several principles of organization: by observable trauma on the body, both outside (bruises and wounds) and inside (traces on the
bones), by agent of death (cuts, blows by various objects, poison, illness, etc.), by social
cause of death (solitary accident, brawl, suicide, homicide, faked causes), by degree of
uncertainty (the *yinan* category, and questions like whether wounds have been inflicted
before or after the death), and so forth." Will, "Developing Forensic Knowledge through
Cases in the Qing Dynasty," 79.

chapters of Song Ci's original *Collected Writings* were maintained in later editions of the *Washing Away of Wrongs*, others were consolidated or otherwise reorganized. Generally, all iterations of these texts contained practical guidelines for examining the body in order to establish cause of death and what in modern parlance is called "manner of death" – that is, the legally oriented classification of whether a death was due to homicide, suicide, accident, or natural causes.[18] These texts also contained sections describing techniques that could be used to treat wounds or revive victims of drowning, hanging, or poisoning, a topic to which the authors of later forensic texts devoted more and more attention. As Xin-zhe Xie and Yi-Li Wu have suggested, officials' use of these techniques to prevent unnecessary deaths was viewed as a measure for improving judicial governance by preventing cases involving injury from becoming homicides in the first place.[19]

Throughout the last century of the Qing, officials, legal specialists, and *wuzuo* critically scrutinized the received knowledge contained in the government-endorsed *Washing Away of Wrongs*, and the errors and ambiguities that they discovered became the focus of detailed treatises that went through numerous editions and continued to circulate in the early decades of the Republic. Various government and private publishers produced editions, often in annotated and expanded form, over the course of the late eighteenth and nineteenth centuries, and important editions, such as that of legal specialist Wang Youhuai, which formed the basis for most of the late Qing expanded editions of the *Washing Away of Wrongs*, circulated widely.[20] Those with experience in forensics also compiled collections of cases that could supplement the official handbook while building a large body of creatively organized knowledge on which officials could draw when confronting situations that had not been addressed in the official literature.[21] These supplementary texts and

[18] For more on the concept of "manner of death" in the contemporary United States, see Timmermans, *Postmortem*, 77.

[19] Xie, "Procedural Aspects of Forensics Viewed through Bureaucratic Literature in Late Imperial China," 2–3; Wu, "Between the Living and the Dead," 58–60.

[20] For discussion of the most important critical editions, see Will, "Developing Forensic Knowledge through Cases in the Qing Dynasty." Republican editions of these texts include, for example, those of Guangyi shuju (1916) and Wenruilou shuju (1921), which reprinted expanded editions of the foundational early–mid nineteenth-century *Buzhu xiyuan lu jizheng*, a text to which was appended a large amount of commentary and cases. The legal medicine expert Lin Ji owned an edition (publisher unclear) of this expanded text, which he continued to use as a reference in his instruction into the late 1940s. Zheng Zhongxuan, "Lin Ji jiaoshou he tade 'Xiyuan lu boyi'" [Professor Lin Ji and his "Critical disputations on the *Washing Away of Wrongs*"], *Fayixue zazhi* 7, no. 4 (1991): 145–8.

[21] Will, "Developing Forensic Knowledge through Cases in the Qing Dynasty"; Xie, "The Shaping of Autopsy Evidence in Nineteenth-Century China."

case collections leant flexibility to the body of knowledge utilized by Qing officials and legal specialists, and established the *Washing Away of Wrongs* as a dynamic site for producing and refining forensic knowledge that was shaped simultaneously by the centralizing interests of bureaucratic procedure and the many insights and experiences of the officials and *wuzuo* who actually examined bodies in local areas. It is important to acknowledge, of course, that beyond the relatively small group of officials and legal specialists who authored these texts, many (if not most) officials would have encountered the *Washing Away of Wrongs* simply as the repository for a set of procedures that had to be followed in order to avoid bureaucratic sanction in the discharge of one's official judicial duties.

In attempting to explain the abundance of medico-legal treatises written in early modern France, Italy, and Germany, Catherine Crawford has argued that the particular fact-finding and evidentiary procedures of Roman-canon law created favorable conditions for physicians to produce a specialist literature on forensics that had no analogue in early modern England.[22] Specifically, Crawford suggests that the codification of forensic procedures and standards of proof in these countries' law codes, as well as the widespread use of written forensic reports as evidence, had the effect of encouraging jurists and physicians to produce a large body of text-based forensic knowledge. This literature took the form of legal commentaries that touched on forensic questions, specialized medico-legal treatises, and collections of expert opinions. There are interesting parallels here with certain forensic developments of the Qing, such as the legal codification of particular body examination practices as well as the attention that authors of legal commentaries paid to explaining forensic procedures and problems.[23] One also finds emphasis placed on the use of highly formalized forensic reports that facilitated judicial review and also provided rich material for forensic treatises and case collections.[24] An important difference, of course, was that in China it was the officials themselves who were supposed to hold absolute authority over the forensic body examination, an

---

[22] Crawford, "Legalizing Medicine."

[23] See, for example, the important role that the Qing Code, legal commentaries, and forensic treatises played in defining Qing practices of wound examination, discussed in Asen, "Vital Spots, Mortal Wounds, and Forensic Practice." Compare with Crawford, "Legalizing Medicine," 99–100.

[24] For the format of the examination reports that were used for much of the Qing and Republican periods, see discussion below. For the use of cases in late imperial forensic works more generally, see Will, "Developing Forensic Knowledge through Cases in the Qing Dynasty." See Crawford, "Legalizing Medicine," 101–3, for the role that written forensic reports played in facilitating the judicial and medical review of evidence by higher authorities in early modern France.

assumption that ultimately distanced physicians from the process of homicide investigation.[25] Indeed, the sophisticated infrastructure of bureaucratically defined examination procedures, tiered judicial review, and instructional texts such as the *Washing Away of Wrongs* was meant to consolidate the authority of officials who did not necessarily have specialist knowledge or experience in this crucial judicial task.

If in China it was supposed to be the bureaucratic officials who maintained authority over the practice of forensic examination, this arrangement relied on the use of subordinate forensic functionaries called *wuzuo*, who were the ones to actually touch the body. Much less is known about those who filled the position of *wuzuo* than about the officials and legal specialists who also took part in these examinations and who authored handbooks or treatises on forensics.[26] There are indications that this position grew out of traditions of corvée labor, in which officials tasked local undertakers with examining the corpse for forensic purposes, a practice that must have been common by the mid-thirteenth century.[27] In the late 1720s, the Qing state issued regulations formalizing this position as part of the organization of local government offices, set quotas for the numbers of such individuals who were supposed to serve in local areas, and established basic procedures for them to study the *Washing Away of Wrongs* with the assistance of local clerks.[28] While these regulations (along with laws issued in subsequent decades) did formalize many aspects of the training, evaluation, and disciplining of *wuzuo* on paper, one gets the impression that provincial officials were often dissatisfied with the knowledge and skills of these individuals, a concern that became part of the justification for forensic reforms that were carried out during the New Policies reform period.[29] Despite the poor reputation of *wuzuo* and officials' often-voiced

---

[25] At the same time, as Silvia De Renzi shows, prosecutors and judges in medieval and early modern Rome did at times deliberate over the meaning of forensic body evidence rather than completely defer to outside experts. Moreover, some legal commentators argued that judicial officials should possess enough medical knowledge to be able to check the evidence submitted by physicians or other experts, a notion that would have been familiar to officials in China during this period as well. De Renzi, "Witnesses of the body," 225, 227; Silvia De Renzi, "Medical Expertise, Bodies, and the Law in Early Modern Courts," *Isis* 98, no. 2 (2007): 320.

[26] Will, "Developing Forensic Knowledge through Cases in the Qing Dynasty"; Li Chen, "Legal Specialists and Judicial Administration in Late Imperial China, 1651–1911," *Late Imperial China* 33, no. 1 (2012): 1–54.

[27] Chang, "'Zhongguo chuantong fayixue' de zhishi xingge yu caozuo mailuo," 12.

[28] *Qinding Da Qing huidian shili, juan* 851, 810.366.

[29] See, for example, Memorial of Xiliang and Cheng Dequan, 8/15/XT1 (1909), FHA 4-1-35-1091-28; Supplementary memorial of Shen Bingkun, 10/7/XT1 (1909), FHA 4-1-38-200-1; Supplementary memorial of Feng Rukui, 5/12/XT2 (1910), FHA 4-1-1-1117-46. For the implementation of the training regulations more generally, see Chen, "'Xiyuan lu' zai Qingdai de liuchuan, yuedu yu yingyong," 66–9.

assumptions about their lack of preparation, it is clear that some became quite experienced and were recognized as "experts" by the authors of forensic case collections and treatises, not to mention the officials who sought their assistance in particularly important or difficult cases.

One of the defining features of the position of *wuzuo* was that those who carried out this work were officially classified as having the status of "mean people" (*jianmin*), a legal classification that made them unable to sit for the civil service examinations or pursue an official career and that carried connotations of moral compromise and unreliability. It should not be surprising that *wuzuo* were defined by this legal and social status, which was shared by other local government functionaries who performed tasks perceived as menial or debased, such as the apprehension of criminals.[30] Only a small minority of people in Qing China had "mean" status, thus differentiating them from the majority of the population which enjoyed the status of being *liangmin*, a word that literally means "good people" but which was used to signify regular commoner status.[31] As is well known, the Yongzheng emperor (r. 1723–35) issued a series of imperial edicts that abrogated the "mean" status of the so-called music households (*yuehu*), a term used to describe individuals who were employed as entertainers and prostitutes as part of a socially marginalized status inherited from their ancestors, as well as a handful of other marginalized groups such as the *duomin* of Zhejiang, who were engaged in a range of occupations.[32] The Yongzheng emperor's reform seems to have been

---

[30] Jing Junjian, *Qingdai shehui de jianmin dengji* [The "mean" social stratum in Qing society] (Hangzhou: Zhejiang renmin chubanshe, 1993), 123–37; Anders Hansson, *Chinese Outcasts: Discrimination and Emancipation in Late Imperial China* (Leiden: Brill, 1996), 48–50. One can speculate as well that the "mean" status of *wuzuo* was related to their routine contact with the dead body, an object that was viewed as "polluting" – physically as well as socially – in other circumstances, i.e., James L. Watson, "Funeral Specialists in Cantonese Society: Pollution, Performance, and Social Hierarchy," in *Death Ritual in Late Imperial and Modern China*, ed. James L. Watson and Evelyn S. Rawski (Berkeley: University of California Press, 1988).

[31] As Matthew Sommer notes, in addition to its usage as a status category, the term *liangmin* also carried connotations of "good" (i.e., moral) behavior, the meaning of the term that is found in legal treatments of chastity and illicit sex during the eighteenth century and after. Matthew H. Sommer, *Sex, Law, and Society in Late Imperial China* (Stanford: Stanford University Press, 2000), 5–8, 71–3, 312–5. For general discussion of these categories and their use during the Qing, see Jing, *Qingdai shehui de jianmin dengji*, 40–8; Sommer, *Sex, Law, and Society in Late Imperial China*, 260–72.

[32] For discussion of the various groups that were affected by the reforms, see Jing, *Qingdai shehui de jianmin dengji*. For the case of the *duomin*, see James H. Cole, "Social Discrimination in Traditional China: The To-Min of Shaohsing," *Journal of the Economic and Social History of the Orient* 25, no. 1 (1982): 100–11. For the Yongzheng emperor's motivations, see Hansson, *Chinese Outcasts*, 163–70, and Sommer, *Sex, Law, and Society in Late Imperial China*, which interprets the reforms as one element of a broader eighteenth-century shift in the organization of sexuality through which obsolete status-based standards of sexual propriety were abandoned.

motivated by a desire for the members of these groups to reclaim a moral standing and subjectivity that they had been denied by association with debased occupations, as well as a more general determination to eliminate punitive social categories that had been put into place by previous dynasties and that had become increasingly irrelevant given the social and economic fluidity of late imperial society. In spite of this shift, however, the *wuzuo* remained an occupational group that was defined by "mean" status until the last decade of the Qing, when officials appealed to the imperial court to abrogate this marginalizing status as a way of reinvigorating and reforming the occupation.

Forensics was thus characterized by officials' simultaneous reliance on and mistrust of the unranked personnel who possessed the necessary specialist knowledge and experience, a pattern that informed local administration under the Qing more generally.[33] *Wuzuo* occupied an ambivalent status given the common perception that they were untrustworthy and willing to subvert the integrity of a forensic examination for monetary gain. Handbooks of forensic examination advised officials on various techniques that could be used to manipulate forensic proceedings. For example, the section on distinguishing real from counterfeit wounds in the official Qing *Washing Away of Wrongs* noted that the *wuzuo* could conceal wounds by dropping madder into the vinegar that would be applied to the body during the examination.[34] The text also noted that when boiling bones, substances might be added to the cauldron that would render the colors on the surfaces of the bones indistinct and make it impossible to find the signs of wounds or poisoning that these examinations were supposed to reveal.[35] The text also included countermeasures that officials could use to detect or counteract these forms of forensic malfeasance. For example, officials could insert a piece of white cloth or paper into the liquids used during the examination to determine if a coloring agent had been added.[36] As these passages suggest, one of the goals of having an official handbook of forensic practice such as the *Washing Away of Wrongs* was to provide officials with a means of evaluating the actions of the forensic examiners who worked under their authority.[37] Ideally, this knowledge was meant to

---

[33] For a critical treatment of the late imperial discourse of yamen functionaries' corruption and a detailed overview of the crucial role that these figures played in local administration, see Bradly Reed, *Talons and Teeth: County Clerks and Runners in the Qing Dynasty* (Stanford: Stanford University Press, 2000).

[34] Lüliguan jiaozheng Xiyuan lu [Records on the washing away of wrongs, edited by the Codification Office] (undated Qianlong edition. *Xuxiu siku quanshu*. Shanghai: Shanghai guji chubanshe, 1995, v. 972), 1.25a.

[35] Ibid., 1.85a.   [36] Ibid., 1.40b.

[37] Chang, "'Zhongguo chuantong fayixue' de zhishi xingge yu caozuo mailuo," 13–14; Will, "Developing Forensic Knowledge through Cases in the Qing Dynasty," 74. For the

serve as a check on the actions of the *wuzuo* and, implicitly, to restrict the possibilities for them to attain autonomous or unregulated control over what happened during a forensic body examination. This was, in broad strokes, the system of forensic examination and the politics of forensic knowledge that the Republican state inherited from the Qing and that it maintained for decades after the fall of the empire.

Remaking the *Wuzuo*

This arrangement, in which officials supervised the work of subordinate corpse examiners, was maintained with slight modification during the New Policies reforms. In late September 1908, the Qing empire's newly created Ministry of Law received orders from the imperial court to deliberate on a memorial submitted by Xu Shichang, Governor-General of the recently established Three Eastern Provinces, and Zhu Jiabao, the first Governor of Jilin.[38] The administrations of Fengtian, Jilin, and Heilongjiang were reorganized in 1907 in order to consolidate Qing control over the Manchu homeland, a region that had recently become an object of Russian and Japanese designs and remained a geopolitically vulnerable part of the empire. This proposal, which had originated in Jilin's provincial justice department, laid out a plan to establish in Jilin's new High Court a school that would enroll literate *wuzuo* as well as youth over the age of 20 *sui* to complete a year-long training program based on the *Washing Away of Wrongs* and Western physiology and anatomy, undoubtedly from Japanese translation. More radically, renamed "inspection clerks" (*jianyan li*), these personnel would become eligible to obtain ranked positions as low-level officials through competitive testing.

The original proposal stemmed from concern over how the forensic capabilities of the Qing empire compared with those of Japan and other countries, which, as the Ministry of Law noted, were based on the employment of physicians trained in Western legal medicine. A key

---

significant amount of forensic knowledge that local administrators would have gleaned from general administrative handbooks, see Xie, "Procedural Aspects of Forensics Viewed through Bureaucratic Literature in Late Imperial China." There were clearly cases in which supervising officials had a grasp of the forensic issues involved and were able to detect the malfeasance of *wuzuo*. For example, for a mid-eighteenth-century case in which a magistrate uses his forensic knowledge to detect a counterfeit wound produced by the *wuzuo*, see *Boan huibian*, 497.

[38] *Daqing fagui daquan* [Compendium of laws and regulations of the Great Qing] (Zhengxue she, n.d.), falü bu, 8.1b–2b. For the administrative reorganization of Manchuria and establishment of the Three Eastern Provinces, see Robert H.G. Lee, *The Manchurian Frontier in Ch'ing History* (Cambridge: Harvard University Press, 1970), 138–81.

concern of the reform was that the forensic personnel of the Qing empire did not measure up to the forensic experts of other countries. The proposal also grew out of long-standing concerns about the distribution and skill of the empire's existing *wuzuo*. By the late nineteenth century, Qing officials were acutely concerned about the difficulties of finding competent examiners and unenforced regulations on their training and local quotas. The idea that raising the status of *wuzuo* would solve these problems was not new. In 1877 the Governor-General of Liangjiang, Shen Baozhen (1820–1879), had suggested that revoking their "mean" status and granting them an occupational status equivalent to clerks of punishments (*xingke shuli*) would improve their quality and encourage more people to become *wuzuo*.[39] While Shen did not discuss the administrative details of this arrangement, granting these individuals the status of clerks would have brought them out of the debased ranks of the runners and given them the status of commoners (*liangmin*), a legal designation that by the late imperial period applied to the majority of people in the empire. Xu Shichang and Zhu Jiabao invoked Shen Baozhen's proposal, which had not been adopted at the time, in their own.

A number of programs for the training of new-style "inspection clerks" were established in the last years of the Qing. In Beijing, the Ministry of Law established a training school under the highest procuratorial office of the central government, the Procuracy General. The students who enrolled in this institution did not include the expected "intelligent youth and literate *wuzuo*," but, rather, low-level officials and degree holders. This prompted yet another proposal to increase the rewards for those who chose to pursue this occupation, in part by replacing the title "inspection clerk" with another title, "inspector" (*jianyan yuan*), in order to further elevate the status of the occupation and accommodate trainees who already held official rank and for whom the lower-status title was unsuitable.[40] Training institutions were established in the provinces as well. In Yunnan, 57 students had completed their training at an earlier school, a second class' instruction was underway, and plans were set for future instruction that would conform to the standards set in the

[39] Wu Yuanbing, ed., *Shen Wensu gong zhengshu* [Official writings of Shen Wensu] (Taipei: Wenhai chubanshe, 1967 [1880]), 7.41a–42b. The connection between yamen functionaries' low status, low "quality," and the difficulty of finding suitable candidates was made in discussions of local administration more generally, i.e., Reed, *Talons and Teeth*, 149.

[40] *Daqing fagui daquan xubian*, falü bu, 6.1a–b. After the fall of the Qing, the judiciary's forensic personnel were generally referred to as "inspection clerks" (*jianyan li*), thus suggesting that this change was not widely implemented. Some forensic personnel did hold the title of "inspector" (*jianyan yuan*) during the 1920s and 1930s, but it was used with much less consistency and frequency than "inspection clerk" during this period.

nationwide reform based on Xu Shichang's proposal.[41] Twelve students had been trained at a smaller program attached to procuratorial institutions in Fengtian and plans were underway for a training institution that could accommodate 100 students.[42] An institution was established in Jilin in late spring 1909, set up with 8 instructors for a capacity of 60 students from localities all across the province.[43] Authorities in Jiangxi were also planning a training program for 100 students from across the province.[44]

The New Policies reform of the *wuzuo* was not an attempt to fundamentally change existing forensic institutions or implement the physician-centered system that Qing observers saw in Japan. This reform maintained the forensic examination of bodies as an activity overseen by bureaucratic officials on the basis of standardized techniques endorsed by the state, not as an area for physicians to apply their own professional expertise. Those who were familiar with the use of physicians in American, European, and Japanese law courts were aware that this reform did not go far enough in adopting what they perceived as the most authoritative forms of modern forensic expertise. For example, Chen Yuan (1880–1971), an anti-Manchu intellectual and early advocate of the Chinese adoption of Western medicine, argued that other countries would still refuse to accept inspection clerks' findings in legal cases that involved their own citizens and that only a deeper utilization of medical expertise in forensics would make it possible to advance China's forensic capabilities.[45] Of course, it is hard to imagine how a physician-centered forensic investigation system of the kind observed in Japan could have been imagined, let alone implemented, at the policy level in China during the last decade of the Qing. The numbers of physicians that would have been necessary to meaningfully reform the Qing state's forensic practices, as well as an infrastructure of medical schools to train forensic experts, simply did not exist at the time.[46]

---

[41] Supplementary memorial of Shen Bingkun, 10/7/XT1 (1909), FHA 4-1-38-200-1.

[42] Memorial of Xiliang and Cheng Dequan, 8/15/XT1 (1909), FHA 4-1-35-1091-28.

[43] Memorial of Chen Zhaochang, 12/17/XT1 (1910), FHA 4-1-38-200-48.

[44] Supplementary memorial of Feng Rukui, 5/12/XT2 (1910), FHA 4-1-1-1117-46.

[45] For an overview of Chen's medical activities, see Andrews, *The Making of Modern Chinese Medicine, 1850–1960*, 118–22. For Chen's writings on forensics, see Chen Yuan, *Chen Yuan zaonian wenji* [A collection of the early writings of Chen Yuan] (Taipei: Zhongyang yanjiuyuan Zhongguo wenzhe yanjiusuo, 1992), 225–37, 250, 284–6, 290–1, 307–8, and 321–2.

[46] By the early 1910s, physicians trained in Western medicine were, in Ka-che Yip's characterization, a "rarity" in China. For estimates of the numbers of Western medicine physicians in China during the first decades of the twentieth century, see Ka-che Yip, *Health and National Reconstruction in Nationalist China: The Development of Modern Health Services, 1928–1937* (Ann Arbor: Association for Asian Studies, Inc., 1995), 13, 132–3, 158–9.

From the perspective of the architects of the Qing empire's judicial reform, it is also unclear whether a more drastic solution would even have been desirable. The system of subordinate body examiners was part of a particular balance of power that gave local magistrates, new-style procurators, and the supervising bureaucrats of the provinces and central government authority over an essential area of judicial practice and the officials and forensic examiners who carried it out in local counties. Delegating forensic examinations to physicians would have lessened the state's control over the technical knowledge that was used while raising new problems of supervision and coordination within a complex judicial system. In the decades following the fall of the Qing, local officials throughout China would be slow to give up the basic assumption that forensic examinations should be handled by subordinate personnel working under the supervision of the state's legal officials. This would have significant implications for the progress of medical reformers who saw forensic investigation as an area of modern governance that demanded scientific knowledge and professional expertise.

Inspection Clerks in Beijing

The work of procurators and their forensic body examiners in Republican Beijing typically began when their office received the report of a death from one of the police districts of the city. Within hours of such a request, a procurator, inspection clerk, secretary, and other judicial officials would proceed to the scene. The jurisdiction of this office included the Inner and Outer Cities as well as the four suburban districts that surrounded Beijing. The considerable work of investigating all of the violent and unexpected deaths of the city was made more harried by the fact that these examinations were carried out at the site of a death, in whichever part of the city it had occurred. As we have seen, Beijing did not have a morgue or other facility to which dead bodies requiring examination or identification could be brought. Police and judicial authorities in nineteenth- and early twentieth-century Paris, by contrast, utilized a series of buildings to store bodies that had been discovered in the city, display them for identification by the public, and support the medico-legal investigations of physicians.[47] Procurators, inspection clerks, and police in Beijing performed the rough equivalent of all of these tasks even

---

[47] For the case of the Paris morgue, see Allan Mitchell, "The Paris Morgue as a Social Institution in the Nineteenth Century," *Francia* 4 (1976): 581–96. For a study of the public visibility of this institution and its status as a spectacular feature of urban everyday life in Paris, see Vanessa R. Schwartz, *Spectacular Realities: Early Mass Culture in Fin-de-siècle Paris* (Berkeley: University of California Press, 1998), 45–88.

as they examined bodies in public and faced the challenge of finishing speedily so that the body could be buried.

The *Morning Post* account of a forensic examination that followed a double homicide in May 1926 can give a sense of what happened once the procuracy's entourage arrived at the scene.[48] In this case, officials were called to a crowded compound located to the east of Xuanwu Gate in the southwest corner of the Inner City. A neighbor had seen a blood-soaked man, who would be identified as a rickshaw puller named Jing Tai, fleeing the residence of Wang Yu, another rickshaw puller, and his wife, née Cao. After the neighbor alerted district police, Jing was almost immediately apprehended. The police soon notified the procuracy and requested that it examine the bodies of Wang and his wife, both of whom had died from the wounds inflicted by Jing Tai. A procurator, Luo Zhenqiu, was sent to oversee the examination, arriving with a secretary, the inspection clerk Zhao Fuhai, a female forensic examiner identified by her surname as Ms. Sun, as well as judicial policemen and other police officers. A little over a month earlier, the same procurator Luo had taken part in the investigation of the March Eighteenth massacre, traveling to PUMC in the days after the shootings to oversee the examination of the body of one of the victims.[49]

Once at the scene of the killing, Luo first interrogated Jing Tai, asking about his relationship with the victims and his reasons for committing the crime. This established that Wang had owed money to Jing Tai and that the two had fought several days before the killing. According to Jing's confession, he went to collect the money from Wang, bringing a dagger that he had purchased after receiving additional threats from him. When Wang attempted to hack at Jing with a knife, he returned the attack, leading to the violence that left Wang and his wife dead. After this round of questioning was concluded, the examination of Wang's body proceeded in the presence of Jing, family, and neighbors. After enumerating the knife wounds on Wang's body, Zhao Fuhai proclaimed to the procurator that they had in fact caused Wang's death, a point already established by questioning. The body of Mrs. Wang née Cao was then examined by Zhao, who enumerated the wounds on the upper part of the body, and then by Ms. Sun, who examined the lower half of the body in another room, out of sight of those attending the inquest. With the investigation concluded, Jing, along with everyone else involved in the case, were released to the procuracy. Thus began the formal stages of preliminary investigation and adjudication before the Capital Local Court.

[48] CB May 10, 1926, 6.
[49] "Youguan sanyiba can'an sishang qingkuang de dang'an," 21.

Who were the individuals who actually examined the body in cases such as this one? By the mid-1920s about a dozen forensic examiners and trainees were working for the Beijing procuracy and its branch offices in the city.[50] Those who served in these positions tended to belong to a small number of clans that provided many of the forensic personnel who worked in Beijing during the Republican period. The importance of the clan is a central theme in the published oral history of Song Qixing, one of the inspection clerks who were trained in Beijing:

The ancestral home of our clan, the Song, is Qihe county, Jinan prefecture, in Shandong. We have pursued the occupation of *wuzuo* since the Ming Dynasty. At the beginning of the Qing, we moved from Shandong to Beijing and still did the old business, serving in the Board of Punishments and passing down this work from generation to generation. By the period of the Beiyang government and Nationalist rule, we served in the Ministry of Justice and later Beiping High Court. After Liberation, one of my nephews and I served in People's Courts in Beijing, Hebei, and Shanxi.

Because the members of our clan had the status of a "mean occupation," the level of cultural attainment was not high, and we relied on the classic work of our occupation, the *Washing Away of Wrongs*, as the theoretical grounds [of our work] while depending on the work experience that has been privately passed on by our forebears to maintain this work as the purview of our clan. Consequently there are no written records of the family history. Now, in discussing the old occupation of "*wuzuo*," I can only rely upon that which was orally transmitted by my elders, and which I myself can recall of what I have experienced since the age of 14 *sui*.[51]

Aside from members of the Song, those of the Yu and Fu (who, according to Song Qixing's account, were relative latecomers to the occupation) also served in these positions. As of the late 1910s, the most senior forensic examiner was Yu Yuan, a highly literate individual who led the training of other inspection clerks and performed skeletal examinations

---

[50] In the late 1910s, there were at least six forensic examiners working for city procuratorial authorities. See the list of names signed on a March 1918 response by Yu Yuan and the other inspection clerks to queries sent by officials in the Ministry of Justice regarding skeletal examination procedures in BMA J174-1-156, 51. A list compiled in May 1922 indicates that ten forensic personnel were employed by the Capital Local Procuracy, with an additional two serving the First Branch office and three serving at the Second Branch. Register of inspection clerks and trainees at Beijing local procuracy and branch offices, May 1922, BMA J181-18-13978.

[51] Song Qixing, "Yitan wuzuo hangdang" [Reminiscing about the occupation of *wuzuo*] in Quanguo zhengxie wenshi ziliao weiyuanhui, *Wenshi ziliao cungao xuanbian: shehui* [Selection of preserved manuscripts of literary and historical materials. Volume 25: Society] (Beijing: Zhongguo wenshi chubanshe, 2002), 395–8. The only information provided about this account was that it was "put in order" by Zhang Gongliang, as dated January 1965. While I have been unable to corroborate many of the details in Song's account, claims pertaining to the Republican era are generally substantiated by court and police files.

for local authorities, an area of forensic practice long acknowledged to be particularly difficult. Yu was recognized as an expert in these examinations and it was not unusual for him to compose written replies to inquiring judicial authorities regarding various points of forensic procedure. Aside from Yu Yuan was Yu Tao, another highly experienced senior examiner, and Yu Tao's son Yu Depei. Two members of the Song, Song Ze and Song Duo, were also forensic personnel at this time. The last was Fu Shun, who had reportedly been the disciple of another member of the Song.[52] Several female forensic examiners who were referred to as "midwives" (wenpo) also worked for procuratorial and police authorities in Beijing. These individuals played an important role in examining the bodies of female victims in forensic cases and investigating wounded living victims and the anonymous dead of the city.[53]

Biographical information about the Beijing procuracy's forensic personnel can be gleaned from documentation surrounding their recruitment and training. Those who received training in a program held by the Beijing procuracy in 1919 brought literacy but learned other skills on the job or as part of their formal training.[54] In their written responses to a brief questionnaire detailing their backgrounds, most enrollees claimed to have studied the Four Books, with some also having read the *Book of Songs* (Shijing) as well as studying other elementary subjects.[55] Information regarding the occupational background of trainees suggests that the majority had been engaged in some kind of commercial enterprise, a background that might have provided another, perhaps more practical, source of literacy.[56] Song Qixing, who went on to serve as an inspection clerk for decades and even after the communist revolution, started out by transcribing judicial documents for the local procuracy and informally following other forensic personnel during their

[52] BMA J174-1-156, 51; Quanguo zhengxie wenshi ziliao weiyuanhui, *Wenshi ziliao cungao xuanbian*, 395.

[53] Song Qixing wrote of them, "At that time the Beijing courts had over 10 inspection clerks. Among them were two 50–60 *sui* female inspection clerks named Ms. Xue and Ms. Wang. Court officials called them 'midwives.' They had been transferred from guard work at the prison, and did not have much knowledge of forensic examination." Quanguo zhengxie wenshi ziliao weiyuanhui, *Wenshi ziliao cungao xuanbian*, 396.

[54] For the training program carried out in 1919, the literacy requirements for enrolled students were that they be "fairly literate and able to write reports." Regulations for enrollment and practical training of inspection clerks, BMA J174-1-27, 6.

[55] These responses are contained in BMA J174-1-67, which includes the enrollees' handwritten responses to questions provided by the procuracy regarding basic personal information, educational and occupational background, and reasons for wanting to become an inspection clerk.

[56] For discussion of the different motives for literacy in late imperial China, see Evelyn Sakakida Rawski, *Education and Popular Literacy in Ch'ing China* (Ann Arbor: The University of Michigan Press, 1979).

examinations.[57] As for their reasons for becoming inspection clerks, several noted that they could "make a living performing forensic examinations" or "provide for my family and make a living," in the words of two recruits.[58] Fu Changling, a 24 *sui* inspection clerk attached to the Capital High Procuracy who joined the training program simply to gain more experience, noted that his forensic training would "not only support my desired livelihood, but could be of some small service to enforcing the country's laws."[59]

Inspection clerks in Republican China would have been a heterogeneous group, shaped by diverse backgrounds, forms of training, and rhythms of work. Some, such as those in Beijing, served under the new procuracies that were established during the New Policies reforms and in the decades after the fall of the Qing. Others served under county magistrates in areas for which modern court institutions had not yet been established or had been decommissioned during the "retrenchment" of judicial institution-building that occurred in 1914 due to a lack of provincial funding to support the ambitious four-level court system that had been planned in the last years of the Qing.[60] The Republican judiciary was itself made up of heterogeneous trial institutions, with very real geographic variations in development, financial resources, and personnel, as Glenn Tiffert has shown.[61] Even during the 1930s, county government officials who were not professionally trained judges or procurators were handling judicial investigation and trial work in about 1,000 of China's 1,500+ counties.[62] In this context, some forensic personnel would have been trained under large-scale programs amid dozens of classmates, whereas others would have received training in small-scale discipleships by following a more senior examiner out into the field. What was typical, though, is that these individuals examined the body as subordinate employees of judicial offices, not as experts in a field of professional knowledge external to the law.[63] Moreover, they did not provide evidence as "expert appraisers" (*jianding ren*), a new legal category roughly

---

[57] BMA J174-1-67, 195.

[58] These answers were given in response to the question "Does the student truly aspire to study forensic examination? Will the student show perseverance? What benefits will there be after completion of studies?"

[59] BMA J174-1-67, 160. Also see Yu Yuan's March 8, 1919 request to the Local Procuracy to have Fu participate in the training program. BMA J174-1-27, 128–30.

[60] Xu, *Trial of Modernity*, 64–5.

[61] Glenn Tiffert, "An Irresistible Inheritance: Republican Judicial Modernization and Its Legacies to the People's Republic of China," *Cross-Currents: East Asian History and Culture Review* 7 (2013): 84–112.

[62] Ibid.; Xu, *Trial of Modernity*, 101–2.

[63] In this sense, there was a significant contrast between the occupational status of inspection clerks and that of the procurators under whom they worked. For more on the high

comparable to the "expert witness" of Anglo-American courtrooms, discussed in Chapter 5. By deploying subordinate body examiners in this way, officials of the Republican judiciary gained a unique form of authority over the dead body, an essential source of forensic evidence in homicide cases.

Inspection Clerks in Action and the Politics of Forensic Knowledge

Compared to the new forms of practical knowledge that literate youth in early twentieth-century urban China could learn in institutions of vocational training and higher education or from commercially available textbooks, the *Washing Away of Wrongs* was not, in any obvious sense, knowledge that would confer economic value or cultural capital.[64] Beijing inspection clerks' skills in examining bodies did provide the basis for a monthly salary that ranged from 14–20 yuan (avg. 19 yuan) in the early 1920s. Those who had recently completed their training and maintained trainee status earned substantially less, ranging from 4–8 yuan (avg. 7 yuan). For comparison, these individuals' salaries were higher than those of police patrolmen (6–9 yuan) in late 1910s/early 1920s Beijing and jail guards (3–7 yuan) in early 1920s Jiangsu.[65] Nonetheless, the technical knowledge at the heart of their forensic practice was seemingly parochial in its pre-modern origins and without the wide applicability of the modern fields of science that were redefining conceptions of expert knowledge and intellectual authority in numerous occupations during this period. The formal training that Beijing's forensic personnel received took the form of a kind of discipleship in which each of the four senior examiners were to accept two students for instruction in techniques of corpse examination as well as practical training in the field.[66] Song Qixing suggested in his oral history that the latter form of training was an established practice within the clan:

standards of education and professional training that defined the careers of judicial officials during this period, see Xu, *Trial of Modernity*, 61–2 and Glenn D. Tiffert, "The Chinese Judge: From Literatus to Cadre (1906–1949)," in *Knowledge Acts in Modern China*, ed. Robert Culp, Eddy U, and Wen-hsin Yeh (Berkeley: Institute for East Asian Studies Publications, forthcoming).

[64] Wen-hsin Yeh, *Shanghai Splendor: Economic Sentiments and the Making of Modern China, 1843–1949* (Berkeley: University of California Press, 2007).

[65] Register of inspection clerks and trainees at Beijing local procuracy and branch offices, May 1922, BMA J181-18-13978. For police salaries, see Gamble, *Peking*, 79–80; Ng, *Legal Transplantation in Early Twentieth-Century China*, 33; Strand, *Rickshaw Beijing*, 72. For prison guards, see Xu, *Trial of Modernity*, 175.

[66] For draft regulations governing this training program, description of the curriculum, and assessments of the recruits' progress, see documents contained in BMA J174-1-27.

Our clan has for generations let boys, starting from the age of 13 or 14 *sui*, follow the older generation out to the site of the examination of the corpse to come into contact with frightening corpses that had undergone all kinds of different deaths. In my own experience, it was from having frequent contact with the dead from a young age that later on I came to be without any misgivings when performing my work.[67]

The training program begun in 1919 was divided into three 6-month terms, each of which culminated in a written examination. From trainees' testing booklets it is apparent that formal instruction focused on passages drawn from the *Washing Away of Wrongs*.[68] Techniques contained in this text formed the basis for students' understanding of how to distinguish wounds from natural post-mortem changes in the body, determine which wound had caused death, and distinguish the myriad other bodily signs that they would encounter during forensic examinations. Trainees also studied a technique described in the *Washing Away of Wrongs* for confirming death by poisoning by inserting a silver needle or hairpin into the cadaver and interpreting the colors that appeared on the object once it was removed, a practice that became a target of forensic reformers' criticisms. Two decades later, forensic examiners in Beijing were still being taught passages from the *Washing Away of Wrongs*, even though by this point a much greater emphasis was placed on integrating understandings of bodily structure and function drawn from legal medicine and Western anatomy.[69]

These body examination techniques were meant to facilitate procurators' control over their subordinate forensic examiners' work. In cases such as the murders committed by Jing Tai, for example, inspection clerks were required to use a standardized routine when examining the body. In this area of judicial practice the Republican judiciary drew directly on late imperial precedents. Officials of the Qing bureaucracy had achieved a great degree of uniformity in local officials' investigation of homicide by requiring that they utilize a standardized form for recording wounds.[70] This form contained a checklist of the parts of the body that had to be examined as well as diagrams that visually represented these anatomical locations (Figure 2.1). In 1918 the Ministry of Justice revised and reissued the original Qing corpse examination forms, maintaining them as the core of the routine that Republican inspection clerks were to follow in their examinations.[71] The form itself was a substantial booklet containing pages

[67] Quanguo zhengxie wenshi ziliao weiyuanhui, *Wenshi ziliao cungao xuanbian*, 395.
[68] BMA J174-1-67.
[69] See BMA J174-2-52 for documents, including testing papers, from a training program carried out by the local procuracy in the early 1940s that shows these mixed contents.
[70] Asen, "Vital Spots, Mortal Wounds, and Forensic Practice."
[71] "Banfa yanduanshu jianduanshu bing shangdan geshi ling" [An order on the promulgation of the corpse examination form, skeletal examination form, and wound list], *Sifa ligui*

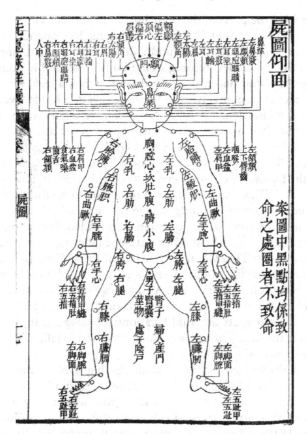

2.1 Frontal view of the body as represented on official Qing examination forms.

with blank spaces in which investigating authorities could write in what they observed on each designated part or, alternately, apply a stamp or other mark to indicate no significant findings. Local authorities in north China directly placed orders for corpse examination forms with Beijing No. 1 Prison, which employed prisoners in printing and other kinds of labor.[72]

Inspection clerks were required to utilize the anatomical terminology of these forms when examining bodies. Much as under the Qing, these

bubian [Supplementary Collection of Judicial Regulations] (Beijing: Sifa gongbao faxingsuo, 1919), 238–61.

[72] Zhuoxian Administrative Office to Capital No. 1 Prison, November 11, 1917, BMA J191-2-13256, 4–6; Wuqiang County Government to Beiping No. 1 Prison, July 7, 1937, BMA J191-2-13739, 7–10; Gamble, *Peking*, 311–2.

documents divided the parts of the body into front and back aspects and "vital" (*zhiming*) and "non-vital" (*bu zhiming*) spots. In conjunction with consideration of the severity of a wound and other factors, late imperial officials had used this understanding of the body's vulnerable points to determine which wound had caused death and to assign legal responsibility for the death in cases involving multiple attackers. The idea was that one could use this knowledge to make systematic decisions about which of several wounds had actually caused death, a crucial question when several assailants had been involved in the affray.[73] Examiners who were trained in Republican Beijing were tested on the anatomical terminology contained in the forms as well as the concepts of wound analysis on which they were based.[74] These anatomical categories also found their way into the deliberations of Republican judges, who routinely relied on the body evidence that the judiciary's forensic personnel provided. For example, in the case of a peddler who fatally stabbed a relative following a dispute that became violent, Judge Tian Chou of the Capital Local Court in Beijing argued that the fact that the wounds were located on "vital spots" demonstrated the assailant's intention to kill. As the judge noted in an exchange with the peddler, the victim "was wounded on the throat and on the abdomen. These two spots are vital locations (*zhiming zhi suo*). If you are saying that at the time you had not already resolved to kill, why choose to attack vital spots (*zhiming chu*)?"[75]

Inspection clerks utilized forensic techniques that they did not create, control, or police. By implication, they could not easily make claims of autonomous authority vis-à-vis the officials under whom they worked. Huang Qinglan, who served as circuit administrator in southern Zhejiang during the late 1910s, admonished county officials under his supervision to familiarize themselves with the *Washing Away of Wrongs*, personally oversee forensic examinations, and "not solely rely on the views of the inspection clerk" – sentiments that would have been familiar to officials during late imperial times as well.[76] It had been a basic goal of the forensic

---

[73] Asen, "Vital Spots, Mortal Wounds, and Forensic Practice," 465–71.

[74] See the responses given on trainees' testing booklets from the first term of their training in 1919. BMA J174-1-67.

[75] CB October 20, 1927, 7; CB November 1, 1927, 7. Also see the Supreme Court's April 1938 reversal of the Hebei High Court's decision in the appellate case of Dong Ping, who had been found guilty of causing the death of Li Sen in a fatal accident that occurred while they were carrying a wooden beam. Jurists of the Supreme Court noted that according to the record of the forensic examination, none of the wounds that Li suffered were on vital spots. On the basis of this observation, and other sources of doubt about whether the wounds had actually caused Li's death, the Supreme Court returned the case to the Hebei High Court. *Sifa gongbao*, no. 8 (1938): Zhongyao caipan, 3a–4b.

[76] Huang Qinglan, *Ouhai guanzheng lu* [A record of governance in Ouhai Circuit] (Taipei: Wenhai chubanshe, n.d. [prefaces 1921]), 3.40b. The use of standardized examination

system implemented by the Qing that officials who were not specialists in forensics should be able to understand the actions of the *wuzuo* and determine whether they were using the officially prescribed knowledge contained in the *Washing Away of Wrongs*. These assumptions about the importance of establishing officials' control over the technical knowledge used in forensic examinations were quite different from those which informed the new conception of professional expertise promoted by physicians of Western medicine, who argued that forensic examiners should have exclusive, autonomous authority in forensics on the basis of their superior knowledge of the body and other physical things. This conception of independent forensic expertise was quite different from the logic underlying the model of bureaucratically controlled forensic knowledge that the Republican judiciary inherited from the Qing.

The Plight of the Procuracy and Its Body Examiners

Relatively few of the cases that the Beijing procuracy handled would have presented twists and turns. One gets the impression from case files and newspaper accounts that fatal accidents, bodies discovered on the street, and even murders and suicides rarely presented mysteries. Often procurators and inspection clerks simply verified facts that were already known, issued a burial permit, and released the body to relatives or the police. The examination protocol that the procuracy's forensic personnel followed, based on the rote examination of a set number of points on the body, was particularly useful for this work. This systematized examination routine provided judicial authorities with a way of closing a large caseload on the basis of what was usually simply an examination of the outer surface of the body. If nothing else, the routine that inspection clerks followed provided a way of stating authoritatively that no marks of violence had been found on a body, a commonsensical standard for ruling out homicide.[77] Given that these examinations were conducted at

and documentary practices also facilitated provincial officials' supervision over subordinate courts, procuracies, and county judicial offices. This paperwork could be evaluated by a province's high court or other regional administrative officials, thus constituting a check on the quality of local forensic examinations. An attentive regional official such as Huang Qinglan could evaluate how carefully local authorities had handled cases on the basis of the standardized forensic examination forms that were submitted as part of the case file. In a number of instances, Huang himself reprimanded local authorities for shoddy work on the basis of these forms' incomplete or inconsistent contents. Ibid., 3.40b, 3.43b, 3.44b, and 3.47b.

[77] That there were "no wounds" on the body was often noted in the newspaper reporting of cases involving deaths from exposure, illness, or other circumstances not related to homicide, indicating the popular currency of this standard, i.e., CB June 11, 1926, 6; CB June 24, 1926, 6; CB January 9, 1928, 7; CB January 13, 1928, 7.

the site of the fatal incident, witnesses and other evidence were made immediately accessible to the procurators who arrived at the scene. While these body examinations did not have the scientific cachet of a physician's autopsy or the sophisticated batteries of testing that could be performed in a forensic laboratory, they were effective nonetheless in the routine work of closing the kinds of cases for which the procuracy was responsible.

The fact that procurators and inspection clerks so often found themselves examining bodies in death cases that were not homicides, and in which no crime had taken place, posed a problem for these officials. As early as May 1914, officials of the Beijing procuracy had attempted to initiate changes in the city's death investigation procedures that would narrow the scope of their forensic examination work. In a letter to police headquarters, the procuracy argued that its forensic examinations should be limited to cases of death and wounding that were actually implicated in crimes or, at least, that actually carried the suspicion of being so.[78] The problem was that the police of Beijing's various districts frequently reported for forensic examination cases of death and wounding that fell out of the "judicial" (sifa) purview of this office. The procuracy often found itself responsible for cases in which someone had died of illness, starvation, or suicide, none of which were criminal cases. Such cases, the procuracy argued, should be handled by "administrative offices" (xingzheng yamen) – that is, the Beijing police – and not its own personnel. While police headquarters acknowledged the legitimacy of the procuracy's concerns in its own reply, without proposals for more specific procedural changes in the cooperation of the procuracy and city police, little came out of the exchange. In fact, police headquarters maintained that its officers should continue to report cases of sudden or unexpected death and suicide in which there was suspicion that a crime had taken place – a point that was not unreasonable, yet that largely undermined the procuracy's call for reform.[79]

It is rather extraordinary that procurators in Beijing were tasked with this work, which was by all accounts considerable, time consuming, and, in many cases, redundant. These forensic investigations became part of their broader professional portfolio, a range of tasks that already covered many stages of judicial process. A lecture on the state of China's procuratorial system given by the accomplished jurist Ye Zaijun (1885–?) at the short-lived Judicial Personnel Training School (Sifa chucaiguan) in early

[78] Capital Local Procuracy to Metropolitan Police Board, May 26, 1914, BMA J181-18-3766, 2–7.
[79] Metropolitan Police Board to Capital Local Procuracy, June 4, 1914, BMA J181-18-3766, 9–11.

1927 touched on the burdens that forensic examinations would have presented to these officials under the best of conditions:

Regarding the current situation of procuracies in China, it is Beijing, Tianjin, Shanghai, and Hankou that are the most booming and populous places. When compared with the rest of the provinces, there are many more cases in these locations. For Beijing procurators, every person accepts on average 10 or more cases every day. A person's energy has limits. One's daily business hours do not exceed seven or eight hours. Forensic inquests are done by him. Questioning witnesses is done by him. Conducting searches and administering penalties is done by him. One still has to appear at court until one is dizzy. How can one avoid being careless?[80]

During the 1920s, some of these burdens were alleviated by judicial police – specially designated officers of the city police force who assisted procurators in investigating crime and managing other aspects of criminal cases. With the order of a procuratorial official in hand, judicial police could be dispatched to oversee the forensic examination in cases of death with little criminal suspicion or involving the examination of wounds on the living.[81] Another measure that lessened procurators' caseload during the 1920s was the police department's own deployment of one or two inspection clerks to investigate the deaths of those who died on the street from natural causes or in poor relief and detention institutions. Under this arrangement, the clerks were dispatched by the Judicial Department of police headquarters, an office that regularly assisted procurators in various matters relating to criminal cases. This arrangement relieved these officials of having to attend forensic examinations, but not the inspection clerks and "midwives" who took on this work as auxiliary to their regular employment at the procuracy.[82]

General Feng Yuxiang's taking of Beijing in October 1924 was an important turning point for the forensic personnel who handled this work. This unexpected move signaled Feng's betrayal of the militarist Wu Peifu (1874–1939) and his Zhili faction, and led to the fracturing of

[80] "Jiancha zhidu cunfei wenti" [On the question of keeping or discarding the procuratorial system], *Sifa chucaiguan jikan*, no. 1 (1927): 99. This institution for the specialized training of judicial personnel began instruction in 1927 and was decommissioned soon after the end of the Northern Expedition. Yu Jiang, "Sifa chucaiguan chukao" [An initial examination of the Judicial Personnel Training School], in *Qinghua faxue* 4, Ershi shiji hanyu wenming faxue yu faxuejia yanjiu zhuanhao [Special issue on jurisprudence and jurists in twentieth-century Chinese civilization], ed. Xu Zhangrun (Beijing: Qinghua daxue chubanshe, 2004).

[81] For forensic examinations overseen by judicial police, see, for example, Report of Inner Left First District, June 22, 1923, BMA J181-18-15363, 12–4; Report of Outer First District, July 20, 1934, BMA J181-20-15724, 2–4.

[82] For Judicial Department documents on the administration of these forensic personnel, see BMA J181-18-18096 and BMA J181-17-2996.

this bloc and a reorganization of military forces and political alliances that ultimately led to Duan Qirui's ascension to Chief Executive in Beijing. Aside from forcing the last Qing emperor Puyi (1906–67) and the remaining members and associates of the Qing imperial family out of the Forbidden City, Feng also disbanded the old Gendarmerie and placed the large regions surrounding the walled city under the direct jurisdiction of the Beijing police.[83] Following this development, the inspection clerks who handled death cases for the Judicial Department were routinely sent into these suburban districts in addition to their regular handling of cases in the Inner and Outer Cities. Over the month of December 1927, for example, Yu Yuan's fifty-one trips for the police brought him to cases of unexpected deaths and discovered bodies in most parts of the city, with the majority located in the Outer City.[84] Cases in the suburban districts could take Yu and his colleagues far beyond the city walls. In early 1928, for example, they examined bodies in the Northern, Southern, Eastern, and Western Suburbs, necessitating round trips of 50–60 li (approx. 17–20 miles).[85] As the individuals who handled these cases were keenly aware, the hot weather of summer months called for examinations to be conducted quickly, a challenge that undoubtedly compounded the difficulties of covering these significant distances.[86] In the end, this arrangement was canceled due to police authorities' unwillingness to pay for the inspection clerks' travel costs. This underfunded area of police operations had long caused financial hardship for the individuals involved, who often had to fight to get their considerable travel expenses reimbursed.[87]

The scope of the city procuracy's forensic examination work did not change significantly until the mid-1930s, for reasons that, interestingly, had to do with the development of a specialized public health infrastructure in the city. In February 1936, the city procurator's office once again

---

[83] For a detailed analysis of the causes and consequences of Feng's coup, see Arthur Waldron, *From War to Nationalism: China's Turning Point, 1924–1925* (Cambridge: Cambridge University Press, 1995), 181–207. For the new administration of the suburban districts by the Beijing police, see Han and Su, *Zhongguo jindai jingcha shi*, 358.

[84] For the list of travel expenses associated with these trips, see BMA J181-20-655. For a list of the same month's travel expenditures compiled by police authorities and also incorporating cases handled by police officials and "midwives," see BMA J181-20-653.

[85] List of inspection clerks' travel expenditures from February 1928, BMA J181-20-668, 62–7.

[86] Song Qiming to Judicial Department, July 1924, BMA J181-18-16827, 3.

[87] On a number of occasions, the forensic examiners' salaries were left unpaid, as were the considerable travel costs required to traverse the city. For the requests that they submitted to the Judicial Department regarding these issues, see BMA J181-18-16827, BMA J181-18-16822, BMA J181-18-19636, and BMA J181-18-19578. For the termination of the arrangement, see BMA J181-20-614, 3–5, which contains an order from 1928 mandating that district police directly contact the procuracy for these cases.

raised the problem of its forensic caseload, imploring the PSB to accept measures that would shift the burden of determining which cases were truly suspicious, and which were not, onto the shoulders of the police.[88] Specifically, it recommended that in cases involving "deaths from illness or unnatural deaths involving no criminal suspicion," police authorities simply conduct an inspection of the body themselves and issue the burial permit on their own authority. If implemented, such a measure would accomplish the goal articulated almost two decades earlier: to limit the scope of this office's forensic examination work to bodies that were actually the focus of homicide investigations. In this instance, as in its earlier proposal for reform, the procuracy's interest was in ensuring a moderate flow of cases that more closely matched its own criminal investigation responsibilities.

It is surprising that the police department in Beijing had not already implemented a mechanism for its own officials to certify unexpected and unnatural deaths without the involvement of the procuracy, especially given the reservations of the latter regarding this issue. The fact that in responding to this proposal the Inner Fifth district head, Wu Kaicheng, proposed that the PSB establish a uniform, city-wide standard for the police-issued burial permits to be used in such cases – thus implying that such a form did not already exist – suggests just how unprecedented this change was for the city police.[89] One underlying factor seems to have been that the Beiping PSB, and the Metropolitan Police Board before it, did not employ personnel who could examine bodies in these cases. While the Beijing police had long relied on yinyang masters to check the bodies of those who died of natural causes, the police districts did not employ their own body examiners who could confirm death from natural causes or collect forensic evidence in homicide cases. The arrangement through which Yu Yuan and other inspection clerks examined bodies with little criminal suspicion under the auspices of police headquarters' Judicial Department was, it seems, as close as the police in Beijing ever came to incorporating this kind of expertise into their own ranks. Yet, even in this instance the police had to rely on an arrangement through which forensic personnel of the procuracy were deployed by them on a part-time basis.

Following the procuracy's proposal in early 1936, internal discussions within the PSB shifted rather quickly to the potential role that the city's new hygiene investigators could play in examining the body in those cases involving little criminal suspicion. Upon receiving word from PSB

[88] Beiping Local Court Procurator's Office to Beiping PSB, February 8, 1936, BMA J181-20-28381, 5–9.
[89] Inner Fifth District to Beiping PSB, February 17, 1936, BMA J181-20-28381, 14.

headquarters that police districts would now be responsible for certifying deaths, Inner Fifth district head Wu proposed that the PSB establish a new arrangement through which the Bureau of Hygiene's statistics investigators and district police examine bodies jointly, thus compensating for the fact that many police personnel lacked the requisite "knowledge of legal medicine" (*fayi zhishi*) to ensure that the inspections were properly done.[90] Within weeks, the PSB and Bureau of Hygiene were drafting procedures through which district police and statistics investigators could cooperate in examining bodies in such cases and these were soon put into place, thus transforming the division of forensic labor in the city by giving hygiene officials a role in investigating cases involving unexpected deaths and discovered bodies.[91] It was under this new arrangement, for example, that when police discovered the body of an unidentified male in an alley of the Inner Third district in early April 1936, they requested that hygiene authorities, not the procuracy, examine the body. The former confirmed what was already clearly the assumption of the police: that it was a case of death from natural causes, with no criminal suspicion at all.[92]

Conclusion

Over the 1920s and 1930s, a very pluralistic arrangement for investigating deaths developed in Beijing. Police, procurators, and hygiene inspectors divided up the work of determining cause of death and explaining the legal, social, and medical implications of death for state and society. Throughout the early Republican period, procurators and their forensic personnel maintained a strong jurisdiction over the bodies of those who died unexpectedly or under suspicious circumstances, an area of work that was guaranteed by the official policies of the judiciary and the institutional practices of the police. Even when it was clear how a death had occurred and that no crime had been committed, the police often persisted in contacting the procuracy for a forensic examination, a practice that effectively established this office's physical control over the dead bodies of city dwellers and its professional authority over forensic investigation work. As physicians of Western scientific medicine made their own new claims over the dead body, they had to negotiate these claims

---

[90] Ibid., 14–15.
[91] Beiping Bureau of Hygiene to Beiping PSB, March 6, 1936, BMA J181-20-28382, 6–11. For the operation of this new division of forensic labor as of the early 1940s, see, for example, "Table of reports of poor people who commit suicide and fall down dead in the Inner Fifth District," appended to report from Inner Fifth District to Beijing Bureau of Social Affairs, March 27, 1943, BMA J2-7-1391, 19–25.
[92] Inner Third District to Beiping PSB, April 3, 1936, BMA J181-20-26463, 9–11.

with those of the procuracy, which never entirely relinquished its authority over the corpses of those who died in unexpected or potentially
suspicious ways. As we will see in Chapter 6, for example, medical schools
that attempted to obtain the unclaimed bodies of prisoners and the
anonymous dead of the city for academic research and the training of
medical students usually had to wait for the procuracy to examine the
bodies before they could be claimed. The growing involvement of city
hygiene authorities in these death investigations beginning in the mid-
1930s did not so much challenge the procuracy's authority over forensic
investigation as narrow the scope of this office's forensic examination
work in ways that more firmly established its jurisdiction over deaths
that were truly suspicious or that were confirmed homicides.

The *Washing Away of Wrongs* and the examination practices that had
been used under the bureaucratic system of the late imperial state thus
played an important role in modern policing and the administration of
justice for decades after the fall of the Qing. These practices were used
across the major turning points that defined this tumultuous period,
whether the warlord politics and militarism of the late 1910s and 1920s,
the deep transformations that followed Beijing's changed status with the
rise of the Nationalist party-state in Nanjing, or the Japanese occupation
that began in 1937 and lasted until the end of the Pacific War. Years
before Chinese medicine was established as an element of China's official
healthcare system in the 1950s, successive Republican regimes were
already relying on a body of late imperial technical knowledge in an
essential area of modern governance. These techniques for examining
bodies and documenting forensic evidence were already highly standardized and easily integrated into the bureaucratic practices of the
Republican judiciary, not only in Beijing but also in local government
offices throughout China. Little modification was needed to make them
compatible with the bureaucratic requirements of a modern state.

In his study of the social organization and practical management of
death in modern Belfast, Lindsay Prior writes:

[Death] is always, in part at least, defined by a whole array of legal, social, religious
and political practices not withstanding the fact that such practices may be histori-
cally and culturally variable. Our comprehension of death and our reactions to
death are therefore constructed in and through the numerous discursive régimes
which circulate around the analysis and examination of human mortality.[93]

Much the same point could be made for early twentieth-century Beijing,
where death became visible and knowable through a multiplicity of

---

[93] Prior, *The Social Organization of Death*, 13.

scientific disciplines and forms of popular and professional knowledge, some of which had historical genealogies that preceded the Republic by centuries. Turning now to inspection clerks' practices of examining skeletal remains in cases involving disputed forensic evidence, we will examine another area of Republican judicial practice in which the forensic techniques of the Qing state continued to play an important role in China's modern judiciary. As forensic personnel such as Yu Yuan were called to re-examine the evidence in cases for which the body had already been interred, they drew on a sophisticated repertoire of techniques for overcoming the epistemological challenges presented by bodily decomposition. These skills in the examination of decomposed or skeletonized remains, which had reached new levels of sophistication under the late imperial state, were co-opted by the Republican judiciary as an essential source of expertise underlying China's modern administration of justice.

3   Disputed Forensics and Skeletal Remains

In late May 1923, officials of the central government's Procuracy General contacted the Beijing procurator's office with the request that it send one of its inspection clerks to Fengtian province to carry out an examination of skeletal remains. An accusation had been made with judicial authorities in Shenyang (Mukden) that a man named Zhao Fukui had kicked a woman named Liu Guangju to death.[1] The aunt of the deceased, a Mrs. Liu née Li, did not accept the findings from officials' earlier examination of Liu's body, which found no wounds and thus exonerated Zhao. Despite the lack of evidence that Liu Guangju had been killed by Zhao Fukui, Mrs. Liu persisted in claiming that her niece had been murdered. She claimed to have seen suspicious discolorations on four of Liu's ribs during the earlier examination, as well as other purple, red, and blue discolorations on the bones.[2] These claims were refuted by the examiners, who affirmed that the bones were pale yellow and without such discolorations, thus providing no evidence of murder. On the basis of her own claims about the forensic evidence, Liu accused the investigating officials of having been bribed to conceal wounds and refused to claim the remains for burial. Authorities in Fengtian now wanted forensic personnel from Beijing to close the case by conducting another examination of Liu's skeletal remains, one that would decisively resolve Mrs. Liu née Li's accusations.

The Beijing procuracy received a number of similar requests from officials in China's northern and northeastern provinces who were faced with the disputed forensics of entrenched cases. These officials valued Beijing inspection clerks' skills in the "steaming" technique for examining skeletal remains, one element in the sophisticated repertoire of forensic practices that the Republican judiciary inherited from the Qing. In response to the request in this case, the Beijing procuracy dispatched Yu Yuan, whom we last encountered training new inspection clerks and

[1] Procuracy General to Capital Local Procuracy, May 23, 1923, BMA J174-1-184, 36–42.
[2] Fengtian High Procuracy to Capital Local Procuracy, June 1923, BMA J174-1-184, 61.

examining bodies for the Metropolitan Police Board, along with Yu Delin to act as assistant. An indication of the reputation of this office, and Yu Yuan in particular, was that Yu himself was called upon to assist in the aftermath of the sensational plundering of the Eastern Mausolea (Dongling) in late summer 1928, an act carried out by the troops of Sun Dianying, an erstwhile commander of militarist Zhang Zongchang who was leading troops in support of the Nationalist cause at the time. Yu Yuan was brought to the site to assist in identifying and collecting the remains of the Qianlong Emperor, Empress Dowager Cixi (1835–1908), and other members of the Qing imperial family that had been scattered amid mud and debris in the devastation caused by Sun's troops.[3]

In early twentieth-century Beijing, the examination of bodies was not usually construed as an activity that could only be accomplished by forensic personnel with special expertise – that is, individuals who possessed knowledge or experience that was recognized to be qualitatively superior to that of the judicial officials who oversaw homicide investigation. "Expertise" and "expert" are used here primarily as analytical categories, not actors' categories, even though, as we will see, there were a range of words in late imperial and early twentieth-century Chinese discourse that were used to articulate notions of authority based on experience, formal learning, or other sources of knowledge.[4] From the cases that the Beijing procuracy handled, as well as administrative documents pertaining to its forensic work, one rarely gets the impression that the status of the individual inspection clerk as an "expert," in this sense, was an important factor for procurators, judges, or others when deliberating on the legitimacy of the forensic findings in this or that case. The most important consideration seems to have been, rather, that the officially endorsed examination procedures had been followed – something that could be accomplished by many of those who were serving as inspection clerks, if not all of them. Distinctions in expertise only became important under a narrower set of circumstances. The examination of skeletal remains, often conducted following an earlier dispute over forensic evidence or during the process of judicial review or appeal, stands out as an area of forensic practice in which late imperial officials and those of the Republican judiciary viewed the expertise of individual examiners as a particularly important consideration for the

---

[3] SJRB August 19, 1928, 3; SNB August 19, 1928, 4; SNB August 21, 1928, 8.

[4] For the history of conceptions of expertise in the early modern and modern West, see Eric H. Ash, *Power, Knowledge, and Expertise in Elizabethan England* (Baltimore: The Johns Hopkins University Press, 2004); Thomas Broman, "The Semblance of Transparency: Expertise as a Social Good and an Ideology in Enlightened Societies," *Osiris* 27, no. 1, Clio Meets Science: The Challenges of History (2012): 188–208.

accomplishment of the task at hand. This was the case because of the procedural importance of such cases to judicial process and the technical challenges involved in examining the bones.

Skeletal Remains and Judicial Process

Judicial officials in late Qing and Republican China faced a challenge that is undoubtedly familiar to those who have examined bodies for forensic purposes in other times and places: the dead body is not a stable object, but one that undergoes physical change – most obviously decay – thus challenging those who would derive usable evidence from it.[5] In the forensic literature associated with the *Washing Away of Wrongs*, one finds a great deal of information about the ways in which the body changes after death, ranging from patterns of discoloration on the skin caused by the post-mortem settling of blood to the physical characteristics and speed of decay under different seasonal conditions.[6] These physical changes were understood to cause various problems for investigating officials. Physical evidence that might be necessary for establishing the cause of death could become obscured, be mistaken for decay or other "natural" post-mortem changes, or disappear entirely along with the flesh. Given that officials and legal specialists understood forensic examinations to be venues in which a range of local actors, including the *wuzuo*, might try to subvert the integrity of the proceedings for personal gain, the problems presented by bodily decomposition were not only "technical" challenges of interpreting evidence, but also raised the very real possibility that an unscrupulous *wuzuo* or false accuser could take advantage of post-mortem physical changes to conceal wounds or assert that there were wounds when there were not.[7]

Managing the potential legal ramifications of bodily decay was thus a clear institutional imperative under the forensic examination system that was put into place under the Qing. The large body of regulations that

---

[5] Ian Burney and Neil Pemberton, "Bruised Witness: Bernard Spilsbury and the Performance of Early Twentieth-Century English Forensic Pathology," *Medical History* 55, no. 1 (2011): 41–60.

[6] See, for example, the chapter on "Distinguishing between the changes of corpses in the different seasons" in *Lüliguan jiaozheng Xiyuan lu*, 1.22a–23a.

[7] This point was made, for example, in Shen Zhiqi's influential commentary on the Qing Code in his discussion of the legal provisions requiring officials to conduct forensic examinations without delay. Shen Zhiqi, *Daqing lü jizhu* [The Great Qing Statutes with compiled commentary] (1746 edition [1715], *Xuxiu siku quanshu*. Shanghai: Shanghai guji chubanshe, 1995, v. 863), 28.51b. Also see Xu Lian, *Xiyuan lu xiangyi* [Detailed explanations of the meaning of the washing away of wrongs] (1854 Preface. 1877 edition of the office of the Provincial Administration Commissioner of Hubei. *Xuxiu siku quanshu*. Shanghai: Shanghai guji chubanshe, 1995, v. 972), 1.30b.

governed forensic examinations, accumulated over the course of the dynasty in statute and substatutes, established penalties for officials who delayed a forensic examination while specifying detailed procedures through which subordinate officials could handle an examination if the county magistrate was unavailable, a measure for avoiding additional delays.[8] The *Washing Away of Wrongs* itself contained numerous passages meant to guide officials in discriminating between "natural" changes that could be expected to occur in the dead body and forensically significant signs such as wounds, and this became a focus of commentaries and appended cases in the late Qing expanded editions of the text. Aside from these measures, one of the most important ways in which Qing officials responded to the problem of bodily decay was to develop knowledge and techniques for examining the skeleton. By the start of the twentieth century, this area of forensic practice was highly developed, a result of Qing innovations that included the promulgation of a standardized terminology for the bones in the 1770s and the significant attention that influential forensic commentators devoted to developing this knowledge over the following century.[9]

There were several different circumstances under which the examination of bones might be necessary. In some cases, a badly decomposed or even skeletonized body might be discovered, prompting officials to focus on the bones from the outset of the investigation. For example, when the skeletal remains of Mrs. Wang née Li, the sister of a brothel owner who was killed for interfering with the plans of one of the prostitutes to escape with a client, were found on the grounds of the Temple of Heaven in October 1927, Yu Yuan and other Beijing inspection clerks carefully examined the remains in a much-publicized "steaming" examination that was covered in the newspapers.[10] In many cases, however, an examination of skeletal remains was only carried out at a later stage of an investigation or trial, at a point when bodily evidence had to be revisited months or even years after the initial investigation. These cases might involve one of several different scenarios. In some cases, local or provincial officials themselves might harbor doubts about the forensic evidence, thus calling for a re-examination of the remains. In other cases, the initial forensic findings might be disputed by a relative of the deceased, thus

[8] *Qinding Da Qing huidian shili, juan* 851, 810.365–77.
[9] Catherine Despeux, "The Body Revealed: The Contribution of Forensic Medicine to Knowledge and Representation of the Skeleton in China," in *Graphics and Text in the Production of Technical Knowledge in China*, ed. Francesca Bray, Vera Dorofeeva-Lichtmann, and Georges Métailié (Leiden: Brill, 2007).
[10] SJRB October 19, 1927, 7; SJRB October 21, 1927, 7; SJRB October 22, 1927, 7; CB February 1, 1928, 7; Zhang Ruilin et al., *Beiping zhentan an* [Cases of Beiping detectives] (Beiping: Wenmei shuzhuang, 1932), 114–20.

leading to an appeal with a higher court. The Qing judicial system had maintained elaborate procedures through which litigants could appeal the judicial decisions made by lower levels of the bureaucracy by bringing the case to provincial authorities or, under some circumstances, the capital.[11] The judicial institutions that were in operation in China in the decades following the collapse of the Qing maintained a sophisticated appellate process, albeit under the auspices of new provincial courts and the Supreme Court in Beijing.

What all of these scenarios had in common was that the administration of justice required that the bodily evidence be revisited at a point weeks, months, or even years after a death had occurred. In the notorious 1870s case of the suspected poisoning of Ge Pinlian by the *juren* degree-holder Yang Naiwu and Ge's wife Bi Xiugu, for example, the forensic inspection of Ge's body that exonerated Yang and Bi was conducted three years after his death.[12] Forensic personnel in early twentieth-century Beijing likewise handled the re-examination of body evidence in cases for which remains had been buried for months, if not years.[13] In such cases, the skeleton became an object of forensic inquiry because of the institutional requirements of the judiciary, which always held out the possibility that an already-interred body would have to serve as evidence later on. If, as we have seen in previous chapters, judicial officials in Beijing routinely made claims over the dead body prior to burial as part of their homicide investigations, skeletal examination cases were ones in which they did so after burial, when exhuming a corpse became necessary to check the accuracy of earlier findings or settle a dispute over evidence. As such, forensic disinterment was another point of contact between the legal imperatives of homicide investigation and the broader world of cultural and ritual meanings surrounding the dead.

The Murky Ethics of Skeletal Examination

In examining the cultural meanings of judicial dismemberment in late imperial China, Brook, Bourgon, and Blue have argued for the existence of broadly held beliefs about the negative religious implications of causing physical harm to the dead body. Specifically, the authors suggest, a mix of ideas drawn from Buddhism, Confucianism, and other elements of popular culture fed the notion that not maintaining the dead body whole would

---

[11] Jonathan K. Ocko, "I'll Take It All the Way to Beijing: Capital Appeals in the Qing," *The Journal of Asian Studies* 47, no. 2 (1988): 291–315.

[12] William P. Alford, "Of Arsenic and Old Laws: Looking Anew at Criminal Justice in Late Imperial China," *California Law Review* 72, no. 6 (1984): 1218–9.

[13] Report of Yu Yuan and other inspection clerks, March 1918, BMA J174-1-156, 44–51.

negatively impact one's experience in the popularly imagined purgatorial afterlife, the form into which one might reincarnate, and the capacity for one's relatives to maintain the ritual bonds between living and dead that formed the bedrock of Cheng-Zhu orthodoxy.[14] In this context, it should not be surprising that the forensic examination of bones could be construed as an act that carried negative implications for the "somatic integrity" of the dead body. According to the legal specialist Shen Zhiqi, who authored an important early eighteenth-century commentary on the Qing Code, if the initial examination of a body was performed well in the first place, "one could avoid the awfulness (can) of a bone steaming examination later on."[15] Shen, like other commentators, did not explicitly address which ritual, religious, or other principles were being transgressed by the procedure, yet it is apparent that if conducted without justification, the examination of skeletal remains left the body in a state that was not normative. Indeed, these examinations required the digging up and opening of the coffin, the exposure of the remains, and an examination process that was, in essence, an extreme form of dismemberment. Each of these was an act that was severely punished under the laws of the Qing state that were meant to protect the integrity of the burial site, coffin, and corpse in order to provide a stable foundation for society's ritual engagements with the dead.[16]

Yet, simply understanding these forensic examination techniques as transgressing cultural and ritual norms surrounding the dead body forecloses other, more sophisticated understandings of how people in China reconciled the social and ritual meanings of the dead body with the legal requirements of homicide investigation and judicial procedure. Officials and *wuzuo* examined bones to obtain the evidence necessary to prove that a death was the result of homicide rather than illness or suicide – an outcome of value to relatives of the victim, and ostensibly even the deceased individual given the belief that suffering an injustice could impact one's experience of the afterlife.[17] Officials could carry out

---

[14] Brook, Bourgon, and Blue, *Death by a Thousand Cuts*, 14–5. As the authors note, these themes are also addressed in Melissa Macauley's discussion of the role that dead bodies played as a powerful symbol of pollution and social ignominy in representations of litigation masters in late imperial China. Macauley, *Social Power and Legal Culture*, 214–8.

[15] Shen, *Daqing lü jizhu*, 28.51b. The examination of skeletal remains was described similarly, as a potentially "awful" (can) treatment of the body meant to be avoided, in the official *Washing Away of Wrongs. Lüliguan jiaozheng Xiyuan lu*, 1.1b. Xu Lian characterized the procedure in the same way in his own commentary on the text. Xu, *Xiyuan lu xiangyi*, 1.32b.

[16] Snyder-Reinke, "Afterlives of the Dead."

[17] That a wronged individual could maintain a presence, if not pursue justice, as a ghost is a notion that appears in literary representations of judicial process more generally in late imperial China. For several examples, as well as a discussion of the complex status of

a skeletal examination in order to reassess and potentially overturn a wrongful conviction, as in the case of Yang Naiwu and Bi Xiugu, thereby preventing a judicial injustice from becoming fixed. In these ways, the decision about whether or not to physically "damage" the dead body by exhuming remains and disarticulating the bones in a forensic re-examination was a complex one, and a question that lay at the intersection of overlapping imperatives surrounding the dead body. Nonetheless, as amply demonstrated by the frequency with which bodily remains were exhumed, scraped of flesh, and boiled or steamed in late imperial times and well into the twentieth century, this was a trade-off that many were willing to make.

One of the most interesting aspects of cases involving the examination of skeletal remains is that relatives of the deceased themselves played a role in deciding the fate of the remains in question – that is, the "steaming" examination was not a compulsory procedure enforced by the state, but one demanded by the kin of the deceased. In the case of Liu Guangju, as we have seen, it was Liu's aunt who pressed for a re-examination of the body on the basis of specific questions about the officials' interpretation of the colors observed on the bones and her accusation that they had been bribed. In refusing to claim Liu Guangju's remains in order to prompt a re-examination of the body evidence, Mrs. Liu née Li was utilizing the corpse as a kind of "tool of legal empowerment," in Melissa Macauley's words, which could keep the case open and force the judiciary to respond to her accusations.[18] In other cases too, the question of whether or not such an examination would be carried out at all was determined by relatives of the deceased, who chose to press the case forward with local or higher-level judicial authorities by refusing to accept the findings of an earlier examination. This reflected the significant responsiveness of the late imperial state and Republican judiciary to the accusations of petitioners and litigants, including those who accused government officials with judicial responsibilities of wrongdoing.[19] The skeletal cases handled by Yu Yuan and other expert

ideas about the supernatural in the practical workings of the Qing judicial system, see Joanna Waley-Cohen, "Politics and the Supernatural in Mid-Qing Legal Culture," *Modern China* 19, no. 3 (1993): 344–9.

[18] Macauley, *Social Power and Legal Culture*, 197.

[19] This point is demonstrated vividly by Quinn Javers, who explores the ways in which officials and common people interacted around cases of unnatural death in late nineteenth-century Ba County. Javers, "The Logic of Lies." For the responsiveness of the Republican judiciary to accusations against local officials launched by ordinary people, see Xu, *Trial of Modernity*, 302–28. For another perspective on the agency of ordinary people in their interactions with the law, see Margaret Kuo's treatment of the new legal regime surrounding marriage and divorce that was put into place by the Nationalist state. Kuo demonstrates that significant new opportunities opened up for regular people,

forensic personnel thus reveal, in an important sense, the commitment of Chinese judicial authorities to re-investigate cases in which the evidence had been questioned.

For another perspective on the ethical and legal issues surrounding the forensic exhumation of bodies as it was negotiated by the state and the relatives of those who died, we might examine the portrayal of the practice in *Cases of Judge Shi* (Shi gong'an), a work of fiction that combined the literary figure of the upright official and judicial themes of Chinese court-case fiction (in this case, inspired by Shi Shilun, a real-life upright official of the early Qing) with the martial heroism and adventurous plots of the late Qing martial arts novel. While the earliest portions of this work seem to have been completed in the last years of the eighteenth century, the earliest extant edition is from the 1820s. A number of other editions, and accretions to the story, were published over the following century, expanding the work into the massive size that characterizes the editions published in the last years of the Qing.[20] Given the work's attention to the details of forensic procedure, it is not surprising that the re-examination of encoffined remains features prominently in Judge Shi's investigations of difficult cases. What is particularly interesting, however, is the way in which the portrayal of the practice in this work also engaged, in indirect ways, with the ethico-legal issues that were at stake in forensic exhumation. For example, in one case that winds through several chapters of the book, a Mrs. Wu née He murdered her husband, Wu Qiren, with a hidden needle stuck in his abdomen, having had an adulterous sexual affair with another man. Throughout the investigation into Wu's death, Mrs. Wu is repeatedly found expressing concern for the treatment of her husband's dead body, an irony that presumably would not have been lost on readers who understood from the beginning that she had instigated the killing.

When the magistrate of Shanyang interrogates her as to why she buried the body of her husband so hurriedly, Mrs. Wu née He replies that a speedy burial is in accordance with both the "laws of the country" (*guofa*), which prohibited the practice of delayed burial, and "human sentiment" (*renqing*), which unquestionably would have endorsed putting her deceased husband to rest.[21] Dissatisfied with her claims of innocence,

including women, to use the law to pursue their own interests. Margaret Kuo, *Intolerable Cruelty: Marriage, Law, and Society in Early Twentieth-Century China* (Lanham: Rowman & Littlefield Publishers, Inc., 2012).

[20] *Shi gong'an* [Cases of Judge Shi] (Beijing: Baowentang shudian, 1982); Meng Liye, *Zhongguo gong'an xiaoshuo yishu fazhan shi* (A history of the development of the art of court-case fiction in China) (Beijing: Jingguan jiaoyu chubanshe, 1996), 123–34; Margaret B. Wan, *Green Peony and the Rise of the Chinese Martial Arts Novel* (Albany: State University of New York Press, 2009), 10–12.

[21] *Shi gong'an*, 985–6.

the magistrate notified her that the investigation would continue with her husband's coffin opened and his body examined, to which Mrs. Wu née He replied:

Given that you, sir, want to open the coffin and examine the body, how can this widow dare to not comply? There is just one thing: if you find wounds on the body, I will willingly plead guilty. If no wounds are found, however, then given that you, sir, have disturbed the corpse without cause, will there not be a penalty for this in the statutes and substatutes?[22]

As Mrs. Wu understood, the evidence was well hidden, and in the end this examination turned up nothing. Soon after the failed examination, however, Judge Shi discovered his own evidence of her role in the murder in the form of an eyewitness to the killing, uncovered after he himself sought out information from local people while disguised as a cloth salesman. When confronted by Judge Shi, Mrs. Wu née He still refused to confess – an act that, as Judge Shi noted, would have its own implications for the body: "If you confess truthfully regarding how you plotted his death, the corpse will not have to be disturbed again, and will not have to lie exposed repeatedly."[23] The subsequent examination showed conclusively that the husband had died from being punctured with a needle, and not from a sudden illness, as had been initially claimed. Mrs. Wu née He persisted in claiming innocence right up until the eyewitness identified her as the one who had killed her husband with this object. In the end, Judge Shi recommended a sentence of judicial dismemberment.

It is not surprising that Judge Shi would want to exhume bodies as part of his investigations. Opening the coffin and removing the remains to examine them was a necessity in the administration of justice, both in the narratives of Judge Shi's handling of cases and in the actual procedures through which late imperial officials investigated homicide cases. Under these conditions, opening the coffin was not necessarily an immoral or transgressive act: the fact that it was Judge Shi, an upright official, who instigated the opening of the coffin suggests the legitimacy of treating the victim's dead body in this way, at least from the standpoint of the story's narrative. Likewise, that it was the murderer who spoke words that, on their face, asserted the importance of protecting the integrity of the body at the expense of the investigation casts doubt on the ethico-legal appropriateness of emphasizing the "somatic integrity" of the dead body above all else at this particular moment in the case. While *Cases of Judge Shi* was the product of particular literary and narrative conventions and cannot be taken as a transparent representation of forensic practice, the work

---

[22] Ibid., 986.   [23] Ibid., 995.

suggests, nonetheless, a basic insight that is crucial for understanding the practice of skeletal examination in late Qing and Republican China: sometimes it was necessary to physically damage a dead body to discover the truth of a crime and pursue justice.

Examining Bones

The first step when examining skeletal remains was to open the coffin and remove the body – acts that would have constituted crimes in late imperial China if carried out under other, unsanctioned circumstances. Any clothing still on the corpse would be removed or cut off and the examiners would cut up the body.[24] At this point, the remains would be cleaned in preparation for steaming. This was accomplished by boiling the body parts in water for two hours and then manually removing any remaining tissue with a knife or other implement.[25] All of these actions were carried out in the presence of judicial officials, relatives of the deceased person, and, depending on the site of the examination, other onlookers. Such examinations would have constituted an unusual spectacle for those who were at the scene. As noted in the account of one examination that appeared in the newspaper *Shenbao*, for example, the inspection clerks "cut away the skin and flesh and dug out the bones, butchering inch by inch, and making it unbearable to watch."[26] As described in the newspaper *World Daily*'s account of the preparation of decomposed remains preceding a forensic examination, "the foul smell swelled up and people were nauseated."[27]

Following these initial preparations, the bones would be placed into a food steamer; in their own examinations, the Beijing procuracy's forensic personnel preferred the use of a large model with three trays.[28] In some instances, the examiners might pack distiller's grains on top of the bones, pouring on wine later on; in others, wine and vinegar might be poured on the bones, which were packed in a porridge made from millet, which was included on the list of necessary supplies that the Beijing procuracy sent to local officials.[29] The application of distiller's grains and vinegar was a common technique in the *Washing Away of Wrongs*, used when

[24] SJRB April 28, 1935, 8.   [25] SJRB October 22, 1927, 7; SJRB April 28, 1935, 8.
[26] SNB February 21, 1930, 15.   [27] SJRB April 28, 1935, 8.
[28] See list of supplies appended to letter from procurator's office of Beiping Local Court to Xinle County Government, September 6, 1929, BMA J174-2-279, 94.
[29] For the use of distiller's grains in these examinations, see *Xiyuan lu xiangyi*, 1.84a. For a case in which the former practice was used, see Chu Minyi and Song Xingcun, "Chi zhenggu zhi huangmiu" [Denouncing the absurdity of steaming bones], *Yiyao pinglun*, no. 36 (1930): 3. For a case involving the latter, see SJRB October 22, 1927, 7. For a case in which Beijing inspection clerks used millet porridge alongside distiller's grains to cover the body, see CB May 7, 1924, 6.

examining bodies for which post-mortem discoloration had made wounds difficult to discern as well as for the examination of bones.[30] These substances were not simply applied to clean the body parts, a task that could be accomplished with water and pods of the soap-bean tree (*zaojiao, zaojia*), a natural substance that had long been used as a cleaning agent in China.[31] Nor was the liberal application of vinegar in such cases simply used to counteract the stench of rotting corpses, even though this might have been understood as an additional benefit. Song Ci's original *Collected Writings*, for example, had instructed officials to use the steam produced by sprinkling vinegar onto heated charcoal to disperse "putrid fumes" (*huiqi*).[32] Rather, officials' use of these materials reflected a long-standing and basic assumption in the forensic literature, simply taken for granted without explanation, that their application would make the physical traces of wounds more visible in flesh and bone.[33]

After these materials were added, the steamer would be placed over a cauldron of boiling water and left for several hours.[34] While the *Washing Away of Wrongs* described two methods for preparing bones – steaming them with wine and vinegar in a pit, or placing them in a large jar with heated vinegar – by the end of the Qing it was a version of the former, different in practice from that described in the text, that seems to have

---

[30] These materials could be applied as part of a forensic examination procedure used to examine bodies that had not decomposed, yet had discolored to the extent that wounds had become indistinct. The process, referred to as "washing and covering" (*xiyan*), involved several steps, which were described in the official *Washing Away of Wrongs* in passages that can be traced back to the thirteenth-century *Collected Writings* of Song Ci. First, the body was flushed thoroughly with water and scrubbed with soap-bean pods. Next, the surface of the body was packed over with distiller's grains and vinegar, covered in clothing doused with hot vinegar, and then wrapped in matting. These actions were meant to apply wet heat to the body for the purposes of rendering it "soft" (*touruan*), a change in physical state that would make it easier to detect wounds on the body surface. *Lüliguan jiaozheng Xiyuan lu*, 1.18a–19a; *Song tixing Xiyuan jilu* [Collected writings on the washing away of wrongs of Judicial Commissioner Song] (Yuan edition. *Xuxiu siku quanshu*. Shanghai: Shanghai guji chubanshe, 1995, v. 972), 1.4a–b, 2.6b–7a. The Beijing procuracy's forensic examiners used this technique on a number of occasions. Report of Yu Yuan and other inspection clerks, March 1918, BMA J174-1-156, 44–5.

[31] In fact, the official *Washing Away of Wrongs* instructed officials to clean the body in this way prior to applying distiller's grains and vinegar. *Lüliguan jiaozheng Xiyuan lu*, 1.18a. For the use of Gleditsia sinensis as a detergent, see Sivin, ed., *Science and Civilisation in China*, 86–8.

[32] *Song tixing Xiyuan jilu*, 2.3b.

[33] For a statement of the power of distiller's grains and vinegar to enhance the visibility of wounds on corpses, see, for example, Xu, *Xiyuan lu xiangyi*, 1.30a–b. One can speculate that applying acidic substances to the bones in this way might have begun to demineralize them, thus making wounds or other forensically significant signs more visible to the naked eye.

[34] SJRB October 22, 1927, 7.

become the more common method.[35] The use of a cauldron filled with boiling water to steam the bones was described by the experienced official and expert in forensics Xu Lian, who wrote an important mid-nineteenth-century critical edition of the *Washing Away of Wrongs*, as one element of the "steaming examinations of recent times" (*jinshi zhengjian*), which differed from the procedure described in the official *Washing Away of Wrongs*.[36] In the official text, the steaming action was achieved by laying the bones inside a hot pit, into which wine and vinegar were poured. By contrast, in the modified technique described by Xu, the remains would be placed on bamboo matting and installed over the cauldron, with distiller's grains added on top of white cloth covering the remains. Using a cauldron of boiling water to produce steam, rather than a heated pit, might have given the examiners more control over the supply of steam. As long as wood or other fuel was added underneath, the steam produced by the boiling water in the cauldron would continue at full force; by contrast, a heated pit would naturally cool.

Upon removal, the bones would be laid out on top of "eight-immortals tables" (*baxian zhuo*), a square table used for everyday purposes, or another surface. The bones were now ready to be examined according to principles included in the *Washing Away of Wrongs*. In order to examine what this process involved, we might draw on the detailed description of a skeletal examination conducted by officials of the Shanghai Local Court in February 1930 that was published by Chu Minyi and Song Xingcun, who advocated the use of Western scientific medicine in legal cases. This account, published by Chu and Song in the journal *Medical Review* (Yiyao pinglun), was based on the coverage of various newspapers and was presented by the authors in order to demonstrate the "absurdity" of these examination practices.[37] The case involved a Mrs. Chen née Sun, whose death had been investigated by authorities of the Wu County Local Court. The forensic examination carried out by this office found that she had hanged herself, despite a number of questionable light wounds on her body, and the outcome of this initial investigation was subsequently appealed by Chen's father. Eventually, the case was given over to the jurisdiction of officials in Shanghai, who were tasked with carrying out a skeletal examination to find the true cause of death. The inspection clerks who handled this case first cleaned the bones of flesh and then placed them in a steamer filled with distiller's grains, later pouring on "burnt wine" (*shaojiu*), a distilled alcohol. Once the bones had been

---

[35] *Lüliguan jiaozheng Xiyuan lu*, 1.32a–b. These instructions were based on passages drawn from the *Collected Writings* of Song Ci. *Song tixing Xiyuan jilu*, 3.2a–b. Also see Xu, *Xiyuan lu xiangyi*, 1.84a.

[36] *Xiyuan lu xiangyi*, 1.84a.   [37] Chu and Song, "Chi zhenggu zhi huangmiu."

removed from the steamer and laid out on a mat, they began to examine them.

The goal of the examination was to determine whether Mrs. Chen née Sun had hanged herself or been strangled to death. Following a process of forensic reasoning that would have been familiar to late imperial officials and *wuzuo*, the inspection clerks who examined the body in this case made a number of claims about the signs detected on the bones that supported the finding that the death was a suicide. First, they discovered a discoloration caused by blood that had seeped into the bone at the left "base-of-the-ear bone" (*ergen gu*), a mandatory examination point on the official Qing skeletal forms that was located on the skull behind the ear; this was the point at which the noose had seemingly come into contact with Mrs. Chen née Sun's body.[38] Their claim that this finding indicated that death had been caused by hanging was consistent with late Qing forensic commentaries, which established that finding a blood-colored mark on this point, or on bones located in proximity to it, could be taken as positive evidence of hanging.[39] They also found it significant that there was no discoloration of bones at the nape of the neck or, as much as could be determined from the flesh that remained on the body, on the skin at this location, a sure sign of the damage that would have been caused by an ever-tightening ligature had Mrs. Chen née Sun been struggling against an attacker. This finding too was consistent with commentary in the *Washing Away of Wrongs*, which claimed that one might find wounds at this location in cases involving strangulation.[40]

Aside from these points, which addressed the direct physical evidence of strangulation on the body, the inspection clerks also made claims about indirect traces, observable on the bones, which could prove that this was a case of suicide by hanging. It was significant, for example, that no discoloration was observed on the skull near the upper part of the forehead, an area of the body that would not have been damaged by a rope or other ligature, yet that could reveal signs of strangulation nonetheless. This was the case because of notions contained in the *Washing Away of Wrongs* about the indirect signs that could be observed on the bones. In cases involving forced strangulation, for example, the *qi* within the body of the victim could be expected to surge upwards, causing a wound-

---

[38] Ibid., 3.

[39] For example, see commentary in *Chongkan buzhu Xiyuan lu jizheng* [Records on the washing away of wrongs with collected evidence, with supplements and annotation, reprinted] (Yuedong shengshu, 1865), 1.45b, as well as Lang Jinqi, *Jianyan hecan* [References for forensic examination], appended to *Buzhu xiyuan lu jizheng* [Records on the washing away of wrongs with collected evidence, with supplements and annotation] (Beizhi wenchanghui, 1904), 6 zhong.11b.

[40] *Chongkan buzhu Xiyuan lu jizheng*, 1.45b.

like discoloration that could be taken as evidence of homicide (the *Washing Away of Wrongs* instructed examiners to look for a small eruption in the skull on the basis of a similar principle).[41] This explanation, absurdly fantastical to Chu Minyi and Song Xingcun, appears less so when viewed within the context of the *Washing Away of Wrongs* and its commentaries, which included other techniques for tracking movements of blood and *qi* in the body by way of traces left on the bones. Such techniques provided the only way to establish cause of death when the flesh had decomposed and all that was left was the skeleton. For example, in the official *Washing Away of Wrongs* one finds the notion that reddish discolorations observed in the roots of the teeth or on the front of the skull could be taken as proof that a victim had sustained a fatal injury to the scrotum.[42] According to the explanation provided by an "old clerk" (*laoli*), cited in the *Washing Away of Wrongs*, regarding the case of a man who died after his murderous wife clutched his scrotum and damaged it, the red discoloration observed on the front of the victim's skull was the result of an upward surge of blood that occurred upon receiving this injury, while the victim's teeth all sustained damage and fell out due to the victim's pained gnashing.[43]

As the authors of late imperial forensic texts were aware, these examinations could involve a great deal of ambiguity as examiners grappled with the myriad variations in condition, color, and evidentiary signs that bones presented.[44] These cases required that an examiner reconstruct what happened around the time of a death on the basis of the limited number of blood-traces and other signs that were observable on bones. A *wuzuo* faced with such an examination might have to account for lost bone fragments and incomplete remains, trace old movements of blood and *qi*, and reconcile the bones in the case with the Qing state's officially endorsed forms for describing skeletal remains, which were commonly acknowledged to be incomplete and inaccurate. As Pierre-Étienne Will

---

[41] Chu and Song, "Chi zhenggu zhi huangmiu," 3; *Chongkan buzhu Xiyuan lu jizheng*, 3.1b–2a.

[42] For an early formulation of this idea, see the late seventeenth-century commentary of Wang Mingde, which was included in the official Qing *Washing Away of Wrongs* of the mid-eighteenth century. Wang Mingde, *Dulü peixi* [A bodkin for untangling difficulties when reading the Code] (Beijing: Falü chubanshe, 2001 [1674]), 8 shang.329; *Chongkan buzhu Xiyuan lu jizheng*, 2.14a–15a. For more on fatal testicular injury in late imperial forensic texts and cases, see Matthew H. Sommer, "Some Problems with Corpses: Standards of Validity in Qing Homicide Cases" (Paper prepared for "Standards of Validity in Late Imperial China," Cluster of Excellence: Asia and Europe in a Global Context, Heidelberg University, October 2013), 38–47.

[43] *Lüliguan jiaozheng Xiyuan lu*, 2.6b. Also see discussion of this case in Sommer, "Some Problems with Corpses," 42.

[44] For some of the challenges, see Xu, *Xiyuan lu xiangyi*, 1.86b–87a.

and Catherine Despeux have shown, the latter was a problem that occupied the officials and legal specialists who authored forensic treatises, and
one suspects that *wuzuo* were no less engaged in the ongoing attempt to
work around the limitations of the official knowledge of the skeleton.[45]
Given the various challenges of the task, it is not surprising that the
examination of skeletal remains was so often associated with the deployment of expert personnel and that some, such as Yu Yuan, came to be
recognized as authorities in this area of forensic practice.

Ambivalent Experts

In late imperial China, the examination of skeletal remains was commonly
discussed among officials as an area of forensic practice that could only be
handled by particular *wuzuo*. When seeking examiners for cases involving
skeletal remains, it was conventional for officials to request *wuzuo* who were
"skilled" (*anlian*) by virtue of experience, a term that appears often in the
administrative discourse surrounding forensics.[46] In late imperial times,
these *wuzuo* could be transferred from other counties and even provinces to
assist in these kinds of examinations.[47] During his time as Surveillance
Commissioner in Jiangsu in the mid-1820s, Lin Zexu (1785–1850) urged
provincial officials to train additional *wuzuo*, noting that a small number of
experienced individuals were repeatedly tapped for the province's skeletal
examination cases:

There are many cases involving loss of life in Jiangsu and oftentimes there are
cases involving examination of skeletal remains, but *wuzuo* who are particularly
adept are very few in number. Several times now, in cases involving the opening of
a coffin and examination of remains, officials compete to summon the *wuzuo* Jing
Qikun of Dantu. It is truly astounding that in a province this big, forensic

[45] Despeux, "The Body Revealed," 644–5; Will, "Developing Forensic Knowledge through
Cases in the Qing Dynasty," 90–1.
[46] For example, Wang Youhuai mentioned the practice in his *Important Points for Handling
Cases* (Ban'an yaolüe, n.d.). Zhang Tingxiang, *Rumu xuzhi wuzhong* [Five works on the
essentials of entering the muyou profession] (1892 Zhejiang shuju edition. Taipei:
Wenhai chubanshe, 1968), 35b. Also see discussion of the practice in *Essentials of
Trying Lawsuits* (Tingsong qieyao), judicial precepts included in the 1890–1 regulations
(nie, guangxu 17, 5a) appended to *Jiangsu shengli sibian* [Provincial regulations of
Jiangsu, fourth collection] (Jiangsu shuju, 1890).
[47] Expanded editions of the *Washing Away of Wrongs* and case collections provide glimpses
of this practice. For example, see the 1791 case of an "old *wuzuo*" named Tang Ming
explaining the skeletal remains of a woman named Zhang Fulian, which was discussed in
the forensic literature as an example of the challenges of skeletal examinations. Tang was
a *wuzuo* from Chenxi county (Chenzhou prefecture), directly northeast of Yuanzhou
prefecture, where the case originated in western Hunan. The brief information included
in the case suggests that Tang was tapped for the case from within the province. *Chongkan
buzhu Xiyuan lu jizheng*, 1.41a–b.

inspections rely exclusively on one person. Given that Jing Qikun is more than 80 *sui*, how can he possibly be employed for long?[48]

By the first decades of the twentieth century, the Beijing procuracy had become an important source for this kind of expertise. We can see in the requests for skeletal examinations received by this office a perception among officials that local forensic examiners lacked experience in such cases and, by implication, that an outside expert was needed. For example, when authorities in Xinle county, southwestern Hebei, sought such an examiner to determine whether a death was due to a fatal beating, strangulation, or suicide, they noted in their request that the local inspection clerk "has scanty learning, is lacking in experience, and is truly not up to the task of examining bones."[49] In May 1923, authorities in Tongxian, a neighboring county to the east of Beijing, requested assistance in two cases involving remains that had decomposed, writing in their request that the county's forensic personnel "lack knowledge of how to examine remains and we fear that they will make a mistake at the critical moment."[50] In another case, sent to the Beiping procuracy in September 1929 by the Liaoning High Court, officials sought out an inspection clerk to handle a skeletal examination case within that province.[51] The case had originated with the procuracy of Liaoyang Local Court, which had contacted Liaoning High Court seeking an experienced examiner for a skeletal case that it was handling. The High Court in turn ordered the Shenyang Local Court to send one of its forensic personnel, yet the individual in question was too ill to participate in the examination. Officials of Liaoning High Court subsequently contacted Beiping authorities to request that they send an inspection clerk who was "skilled at steaming examinations" (*jing yu zhengyan*) to handle this important task. According to the Liaoning officials' request, there was no one else in the office with suitable skills to conduct the examination.

Beyond handling skeletal cases such as these, expert examiners such as Yu Yuan could also provide assistance to judicial officials in the form of specialized information about forensic procedure. In March 1918, for example, officials of the Ministry of Justice contacted the Beijing procuracy with the request that this office question its "veteran inspection

---

[48] Lin Zexu, *Lin Zexu ji: gongdu* [Collected writings of Lin Zexu: Administrative documents] (Beijing: Zhonghua shuju, 1963), *gongdu* 2, 11–12.

[49] Hebei High Court procurator's office to Beiping Local Court procurator's office, July 20, 1929, BMA J174-2-279, 78.

[50] Tongxian Government Office to Capital Local Procuracy, May 19, 1923, BMA J174-1-184, 19–21.

[51] Liaoning High Court procurator's office to Beiping Local Court procurator's office, September 9, 1929, BMA J174-2-279, 127–30.

clerks" (*laolian zhi jianyan li*) regarding questions about skeletal examination procedure that had arisen in a case that had been handled by the High Court of an unspecified province.[52] At the center of the case was the body of an unnamed individual who had died in December 1915. The coffin remained untouched for several months until a relative opened it in February 1916, subsequently lodging an accusation with county authorities that the deceased had been killed by another individual, who had allegedly poured water and kerosene oil down the victim's throat after having employees beat the victim. An inspection of the body carried out by county authorities found that the deceased had died of natural causes, thus refuting the claim that a murder had been committed. Given the relative's repeated and ongoing accusations, however, county officials requested that the High Court send an inspection clerk to re-examine the body.

The behavior of the person who was sent to handle this task raised questions for officials of the Ministry. Upon his arrival at the county, this individual delayed conducting the necessary "steaming" examination because the encoffined body had not fully decomposed. Returning two months later, he once again delayed the examination for the same reason. In the end, it was only two months after this second trip, when the inspection clerk again returned to examine the remains, that the steaming examination was finally carried out. The case had raised three questions for officials of the Ministry, who now contacted the procuracy in Beijing. These were, first, whether there were methods for examining bodies that had begun to decompose but had not completely skeletonized; second, whether one really did have to wait for the complete dissolution of the flesh to carry out a "steaming" examination; and third, whether the examiner in this case had been following proper procedure when he collected teeth and assorted bones that had separated from the decomposing corpse and stored them separately from the coffin.[53]

That the Ministry of Justice sought answers from the Beijing procuracy and its forensic personnel is interesting for several reasons. This exchange raises questions about how officials who were not themselves experts in forensics formulated forensic policy amid an arrangement of specialist knowledge in which those who ostensibly knew the most about this topic were subordinate personnel of lower social status. That the Ministry contacted the Beijing procuracy in this way also suggests the important role that inspection clerks might play as sources of authoritative

---

[52] Criminal Affairs Department of the Ministry of Justice to Capital Local Procuracy, March 23, 1918, BMA J174-1-156, 37–41.
[53] Ibid.

knowledge, and knowledge of the kind not commonly possessed by officials themselves. Some examiners clearly could be trusted and, more than this, could serve as a source of useful insights, a pattern that already had precedents in late imperial times. For example, forensic treatise-writers such as Xu Lian drew on the experience of *wuzuo* when evaluating the validity of various claims about the anatomical structure of the body in the official *Washing Away of Wrongs*.[54] That an official such as Xu would seek the insights of subordinate body examiners in the first place suggests a potential world of exchange between officials and *wuzuo* that belies the conventional image of the latter as categorically corrupt and in need of discipline and supervision.

There were also precedents for *wuzuo* to play a role in the formulation of forensic policy. We find a particularly important example in the Qing state's promulgation of a standardized terminology for examining skeletal remains, a development that occurred about a century and a half earlier. By the late eighteenth century, officials and *wuzuo* were already required to use a standardized checklist of the parts of the body when recording wounds and other bodily signs, a practice that required them to "code" each part of the body as a vital or non-vital spot, a distinction that carried significance when sentencing fatal affrays under the substatutes of the Qing Code. In the intercalary fifth month of the 35th year of the reign of the Qianlong emperor (early July 1770), the Board of Punishments was ordered to deliberate on the proposals of the Anhui Surveillance Commissioner, a Manchu bannerman named Zengfu, who had memorialized on the challenges of examining skeletal remains without a standardized listing of bones that could supplement the discussions of the skeleton in the *Washing Away of Wrongs*.[55] Creating a checklist of the bones, he argued, would better equip officials to handle this challenging area of forensic examination while ensuring that they did not make errors when deciding the statutorily important question of whether a wound had been inflicted on a vital or non-vital spot.

Officials of the Board agreed with this proposal and created a new checklist of the bones and images of the skeleton, modeled on the existing corpse examination checklist and images, which became part of the official *Washing Away of Wrongs*. Despite legal specialists and officials' criticisms of these new documents and subsequent attempts to improve the skeletal knowledge contained within them, they nonetheless constituted a technical standard that officials and *wuzuo* had to follow in their

[54] Xu, *Xiyuan lu xiangyi*, 1.58a, 1.59b, 1.93a.
[55] *Qinding Da Qing huidian shili, juan* 851, 810.374. For a brief biography of Zengfu, see Yin Haijin, *Qingdai jinshi cidian* [A dictionary of *jinshi* degree-holders of the Qing dynasty] (Beijing: Zhongguo wenshi chubanshe, 2004), 342.

examinations.[56] What is particularly interesting about this development is that in formulating these new materials, the Board of Punishments had relied on the assistance of *wuzuo*. Specifically, officials of the Board noted in their description of this process that, aside from selecting veteran officials, they also gathered together "experienced *wuzuo* of various offices" (*ge yamen jingxi wuzuo*) to take part in the process.[57] In this instance, much as in the exchange following the Ministry of Justice's queries in 1918, *wuzuo* were asked to provide specialist knowledge that, in the end, would be used by officials to further consolidate the bureaucracy's authority over subordinate body examiners. As Zengfu himself had implied, having a standardized list of the points on the skeleton that needed to be examined would facilitate officials' supervision over the *wuzuo* who carried out skeletal examinations, an anxiety about the integrity of forensic proceedings that had long motivated officials and legal specialists to disseminate technical knowledge to officials through the *Washing Away of Wrongs* and other texts on forensics. Similarly, the point of the Ministry's communication with the Beijing procuracy in 1918 was to evaluate the actions of the inspection clerk in this case and to determine whether he had followed proper procedures in delaying the examination and storing the remains in the way that he did.

These instances of officials seeking the assistance of subordinate body examiners raise an important question: on what basis did expert *wuzuo* or inspection clerks claim authority? In other words, what were the sources of their expertise? Returning to the exchange from 1918, we find some evidence in the terse document signed by Yu Yuan and the Beijing procuracy's other forensic personnel that addressed the Ministry's questions. The report addressed each of the questions in turn, providing examples of "cases that have been handled" (*banguo cheng'an*) to substantiate the claims that Yu Yuan made for each. For example, in responding to the officials' query about whether one had to wait for the complete skeletonization of the body before beginning a "steaming" examination, Yu wrote that such an examination could be carried out

---

[56] Later authors' critical assessments of these materials can be followed in Despeux, "The Body Revealed" and Will, "Developing Forensic Knowledge through Cases in the Qing Dynasty," 90–1. Interestingly, Yi-Li Wu has found that the Board's checklist and diagrams were used by literati-physicians with an interest in trauma medicine to refine their understandings of the effects of injury on particular parts of the body and supplement their knowledge of anatomical structure. Wu, "Between the Living and the Dead," 64–8.

[57] *Qinding Da Qing huidian shili, juan* 851, 810.374. For more on the process through which the officials created these new visual representations of the bones, see Han Jianping, "Huashe tianzu: Qingdai jiangutu zhong de pianzhi guge" ["Drawing Legs for a Snake": The Superfluous Bones in the Qing Dynasty Bone Inspection Diagrams], *Kexue wenhua pinglun* 8, no. 6 (2011): 58–67.

whenever necessary, regardless of the degree of decomposition. Yu then cited three cases from the last decade of the Qing and early years of the Republic in which bodies were found in states of decomposition ranging from the complete disappearance of the flesh to the partial decay of the body. In responding to the other two queries, Yu cited six other cases in which bodies had been examined in Beijing or around central Hebei that demonstrated the proper procedures for storing, transporting, and examining remains in such cases.

We see in this report Yu Yuan's deployment of a particular strategy for claiming authority: citing cases that could validate his claims about forensic procedure and demonstrate his experience. The compilation and use of such "leading cases" (cheng'an) had long been essential to the process through which late imperial jurists and legal specialists reconciled the statutes and substatutes of the Qing Code, as well as the instructions provided in the *Washing Away of Wrongs*, with legal and forensic scenarios that had not been addressed explicitly in these authoritative texts.[58] By presenting cases in this way, Yu Yuan was invoking an authoritative category of knowledge, one that the officials of the Ministry would have recognized as generally important in the world of law. This was not the only time that Yu Yuan used cases to claim authority in this way: the same strategy was used five years later when he wrote a reply to the queries of authorities in Fengtian who were investigating the case of Liu Guangju.

In this instance, Yu Yuan once again composed a written report explaining the procedures for examining individual bones rather than the entire skeleton – the procedure that officials initially requested in this case. Yu Yuan and his colleagues in Beijing had conducted a number of examinations that focused simply on the one part of the body that was being disputed by relatives of the deceased – for example, the skull or ribs – thus saving the trouble and expense of examining the entire skeleton. When using such a procedure, Yu explained, it was essential to limit the dispute to the one piece of bone at issue and for the relatives of the deceased to complete a bond indicating their willingness to accept the findings of the re-examination. In order to demonstrate this point, Yu cited five cases from across north China with which he had been involved personally:

I have had experience with this method in [the following cases]: Han Tuer of Taizishan area, Zhangjiakou, who was beaten on the frontal eminence and in which only the skull was examined; Gu Chundi's daughter-in-law Mrs. née Wang of Xidayuan area, Tianjin, who was wounded on the face and in which only the skull was examined; [The case of] Mou Chuanzi of Fushan county,

[58] Will, "Developing Forensic Knowledge through Cases in the Qing Dynasty," 64–8.

Shandong accusing Mou Caiting of beating his father on the right ribs and in which only the right ribs were examined; [The case of] Xin Changqing of Wanghuizhuang, Sanhe county reporting that his daughter Mrs. Yan née Xin was wounded on the left rear ribs and in which only the left rear ribs were examined; [The case of] Wu Dianru of Banbidian, Tongxian reporting that his brother Wu Dianxiang was injured when his scrotum was seized, and in which only the skull was examined. Each of the above cases was one in which a relative of the deceased designated the one vital spot (*yaohai chu*) for examination and stated under bond that they were willing to have a steaming examination in which only that bone was examined.[59]

As this passage suggests, what qualified Yu Yuan to explain the process of examining single bones was his prior experience with the procedure. This experience constituted, in this instance, the basis of Yu Yuan's expertise – a source of authority in forensic matters derived from the simple fact that Yu had handled forensic examinations for a long time, had seen variations in the procedures, and knew what was expected. Yu's authority in forensic examinations did not derive from an academic credential or from training in a field of professional knowledge such as medicine that was fundamentally distinct from the working knowledge of the judiciary – an arrangement that ostensibly would have given inspection clerks a greater degree of autonomy in their dealings with officials. Indeed, while Yu might claim authority on the basis of experience, this was hardly a resource that could be the exclusive possession of inspection clerks as an occupational group. Authors of forensic treatises such as Xu Lian and, in Yu Yuan's own time, a Beijing procurator named Wang Chichang who authored an early twentieth-century expansion of the *Washing Away of Wrongs* utilized such cases in their own works, in the process claiming authority on the basis of their own forensic experiences as well as those of others.[60] Despite the fact that inspection clerks were the most immediate claimants to forensic expertise in the sense that it was they who actually handled the body, they could hardly have claimed exclusive authority in forensics given the institutionally mandated role of officials in supervising their examinations.[61] Thus, while distinctions between expert and non-expert might exist, the boundaries between the two were highly porous.

---

[59] Report of Yu Yuan, June 11, 1923, BMA J174-1-184, 71–2.
[60] Asen, "The only options?," 151–2.
[61] For more on this point, see Chang, "'Zhongguo chuantong fayixue' de zhishi xingge yu caozuo mailuo," 14.

Conclusion

Examining how the expert authority of forensic examiners such as Yu Yuan was conceived has revealed another dimension of the particular regime of forensic knowledge that undergirded the judicial system of the Qing and that came to serve the new administrative geography of the Republican state. The work of experts such as Yu Yuan demonstrates the dual-tiered framework in which forensic examinations were carried out. Most commonly, forensic examinations were conducted locally, wherever a homicide case occurred. The underlying principle was that local officials and inspection clerks of unknown skill and experience would utilize easily practicable techniques for examining bodies and documenting the findings, carried out under layers of supervision that would ensure that the proper techniques had been used. In some cases, however, problems might arise following these initial body examinations, whether doubts about the local officials' interpretation of the evidence or even a relative of the deceased disputing the findings. In such instances, it was common for officials to seek the assistance of a smaller group within the forensic personnel of the judiciary, those of recognized expertise in the skeletal examinations that were required when a body had to be re-examined. Under this arrangement, the use of "experts" was less an overriding institutional goal to be applied in every forensic examination than a necessary response to the extraordinary circumstances that might be encountered in a much smaller subset of cases.

Skeletal examination cases thus involved a way of organizing technical knowledge that was very different from that which was used in the most routine forensic examinations. In such cases, the most important source of forensic knowledge was the individual *wuzuo* or inspection clerk who traveled to the locality in which the case had occurred. The mobility of these experts thus looms large as a theme in such cases – indeed, it was the transit of their own bodies to other counties and provinces that made it possible for the necessary technical knowledge to be applied at all.[62] This contrasts with the most common ways in which forensic knowledge was organized in China during this period: through the wide dissemination of easily accessible knowledge contained in the *Washing Away of Wrongs* and other forensic texts. It is in this sense that, much as Francesca Bray has noted for officially sponsored treatises on agronomy, late imperial

---

[62] For the importance of migration and mobility as vectors for the transmission of technical knowledge in the medieval and early modern West, see Liliane Hilaire-Perez and Catherine Verna, "Dissemination of Technical Knowledge in the Middle Ages and the Early Modern Era: New Approaches and Methodological Issues," *Technology and Culture* 47, no. 3 (2006): 559–64.

handbooks of forensic examination were important vehicles for "comprehensive, mobile knowledge that could successfully be transferred through the medium of print, across the vast spaces of the empire, and translated into local action."[63] The use of expert examiners who possessed knowledge beyond that which could be gleaned from the texts implicitly challenged the state's bureaucratic control of forensic knowledge and the mechanisms of oversight and supervision on which the Qing forensic examination system was based.

Skeletal examination cases also demonstrate the great extent to which Chinese forensic practices were integrated into the bureaucratic structures of the state and, specifically, the hierarchically organized judicial infrastructure. Officials of the Republican judiciary relied on networks of bureaucratic communication to seek out skilled examiners from other counties and provinces for these cases. In the case of Liu Guangju, for example, Beijing authorities received the request for assistance from the Procuracy General, which had been contacted about the case by officials of the Fengtian High Procuracy in response to difficulties faced by county-level judicial officials within the province.[64] During the late 1920s, the procurator's office of the Hebei High Court served as a kind of broker in such cases by putting the Beiping procuracy into contact with county authorities in Hebei, and, in at least one case, authorities in Jilin who were seeking an inspection clerk to handle a skeletal examination.[65] In these ways, the judicial bureaucracy provided an important network through which such requests were transmitted. This reflected the significant role that higher-level authorities within the province played in local judicial affairs, both in the formal system of appeals and also as a source of information and advice regarding the forensic evidence of cases that were in the process of investigation. Similar dynamics would inform county-level officials' engagements with China's first legal medicine laboratories during the 1930s, an arrangement that similarly relied on the judiciary's networks of communication to put local officials into contact with new kinds of forensic experts in Beiping and Shanghai.

---

[63] Francesca Bray, "Science, Technique, Technology: Passages between Matter and Knowledge in Imperial Chinese Agriculture," *The British Journal for the History of Science* 41, no. 3 (2008): 334.

[64] Procuracy General to Capital Local Procuracy, May 23, 1923, BMA J174-1-184, 36–42.

[65] For the latter, see Hebei High Court procurator's office to Beiping Local Court procurator's office, April 3, 1929, BMA J174-2-279, 1–8.

4   Publicity, Professionals, and the Cause of Forensic Reform

On March 7, 1928, a body was found near Taiping Lake on a strip of land located between the southwestern edge of the Inner City and the campus of Republican University. As was typical in such cases, the police officers who responded to the discovery of the body requested that the procuracy conduct a forensic examination. Under supervision of one of the city's procurators, an inspection clerk named Liu Qipeng examined the body of the dead male and found wounds on the head, chest, and hands, and a leg that was severely broken, all apparent signs of a fatal beating. As was subsequently revealed in the newspapers, Liu also claimed to have found traces of semen during his inspection of the victim's genitals, a detail that suggested to the officials at the scene that the man's death had been the violent result of the discovery of his illicit sexual affair with an unknown party. Because the body remained unidentified following the forensic examination, district police encoffined the corpse pending further investigation.[1]

The mystery of the corpse discovered at Taiping Lake became more complicated almost overnight. City police soon established connections between this case and an incident that had occurred the previous evening on the other side of town. A city official named Feng Xiangguang had been riding in his private automobile in a part of the Inner City just inside of the Chaoyang Gate when his driver, Jin Dasheng, struck a person on a bicycle. Feng left the scene after ordering Jin to take the injured person to a hospital. Later in the evening, Jin contacted Feng's household, claiming that the rider's injury had not been serious. Police in the area soon found out about the incident and busied themselves with confirming Jin's claim that the injured person, who conveniently happened to be one of his acquaintances, was convalescing. At the same time as these events were unfolding, a woman reported the disappearance of her husband Su Baojie, a bank courier who left home on his bicycle and never returned.

---

[1] For police reports detailing the course of the investigation in this case, see BMA J183-2-3836. Also see CB March 9, 1928, 7; CB March 10, 1928, 7; SJRB March 9, 1928, 7.

A shoe discovered near the scene of Jin Dasheng's automobile incident turned out to be Su's, and upon further investigation by the police it became apparent that it was Su's body that had been found near Taiping Lake. Jin's story rapidly fell apart.

A second examination of the body was conducted on the afternoon of March 8, one day after it had been discovered. As the coverage of both *Morning Post* and *World Daily* made clear, Liu Qipeng had been completely wrong. The wounds that had seemed to have been the result of an assault were now clearly seen as having been caused by the impact of an automobile on the victim's body. Liu's now inexplicable claim that the killing had had a sexual angle likewise appeared to have been profoundly misguided, a mistake that "not only endangered the interests of the victim but also damaged the prestige of the court," as the Beijing Bar Association, the pre-eminent professional association for lawyers in the city, subsequently argued.[2] The second examination of the body was a raucous event, carried out in the presence of masses of spectators, many of whom watched from the bordering walls of Republican University. Following the opening of the coffin, Su's wife identified the body, which was examined again by inspection clerks of the procuracy. Following a city-wide mobilization of police, Jin Dasheng was soon apprehended by detectives. Under the direct orders of the militarist Zhang Zuolin, who had controlled the capital since mid-1926 and presided as Generalissimo of the government since June 1927, Jin was handed over for execution following a dramatic procession through the streets of Beijing that took place a little less than two weeks after the discovery of Su Baojie's body. As described in a sternly worded official statement that was published in the newspapers, the execution was meant to be a warning to those automobile drivers who would injure or kill pedestrians and flee the scene, a disturbing pattern that had emerged as more and more vehicles took to the streets.[3]

The 1920s saw a spate of cases in which errors in forensic evidence were traced back to judicial officials and inspection clerks in the very public venues of newspapers, professional journals, and other printed media. Much as in the killing of Su Baojie, these cases attained great visibility in an urban press that was already providing readers with daily coverage of violent and unexpected deaths and their investigation by city authorities. This coverage did more than simply disseminate the facts of cases such as that of Su Baojie to readers. Rather, the visibility of these cases galvanized

---

[2] Beiping Bar Association to Beiping Local Court and Procuracy, July 31, 1928, BMA J174-2-152, 46.
[3] CB March 18, 1928, 7; SJRB March 18, 1928, 7.

socially engaged individuals and professional groups, including physicians of Western medicine, to criticize the state of China's forensic practices and promote the reform of a forensic system that, in their estimation, relied on outdated techniques.[4] In the process, newspapers, literary supplements, and professional journals became important sites for critical discourse and debate on the integrity of the state's body examination practices and the perceived shortcomings of legal officials and their forensic personnel. In the post-May Fourth moment of the 1920s, the question of forensic reform easily became implicated in pressing issues of the day, such as the progress of Chinese judicial reform and the push for foreign countries to rescind their extraterritorial judicial privileges, as well as discontent with the pace of China's adoption of Western science.[5]

The emergence of this critical discourse on the judiciary's forensic practices had much to do with the public activism of members of the Western medical profession and their use of the printed word as a tool of public engagement and influence, a reflection of the fact that the rise of the modern professions in China coincided with the deepening of industrial print capitalism and the expanding circulation of newspapers, textbooks, and other commercially printed material.[6] Over the course of the nineteenth and early twentieth centuries, many societies saw organized bodies of physicians assert a more active role in the forensic investigation of deaths, a process of medical professionalization that, to varying degrees in different times and places, established a significant role for physicians in the administration of justice. A similar process began in earnest in China during the 1920s as practitioners and proponents of legal medicine began to redefine forensics as a public issue of reform while attempting to persuade legal officials to accept the involvement of physicians in their

---

[4] For other instances in which China's modern print media facilitated the mobilization of social and professional interests surrounding cases of homicide and suicide, see Eugenia Lean, *Public Passions: The Trial of Shi Jianqiao and the Rise of Popular Sympathy in Republican China* (Berkeley: University of California Press, 2007); Madeleine Yue Dong, "Communities and Communication: A Study of the Case of Yang Naiwu, 1873–1877," *Late Imperial China* 16, no. 1 (1995): 79–119; Carroll, "Fate-Bound Mandarin Ducks"; Bryna Goodman, "The New Woman Commits Suicide: The Press, Cultural Memory, and the New Republic," *The Journal of Asian Studies* 64, no. 1 (2005): 67–101.

[5] Much as in late nineteenth- and early twentieth-century Siam, the unequal legal and political relations that were established between China and other countries around extraterritoriality played an important role in driving forensic reformers to argue for a new model of forensic investigation based on medical expertise, the standard that, it was often assumed, was required to match the forensic practices of other countries. Pearson, "Bodies Politic," 199–243.

[6] Christopher Reed, *Gutenberg in Shanghai: Chinese Print Capitalism, 1876–1937* (University of Hawai'i Press, 2004).

cases. Proponents of forensic reform promoted the new idea that professional physicians should hold the exclusive authority to interpret the significance of physical evidence, especially the body. This argument represented not a higher valuation of physical evidence per se – as we have seen, Chinese legal officials already used body evidence in sophisticated ways – but, rather, the new assumption that physicians' knowledge of the body and other physical evidence was more accurate and authoritative than that of legal officials or other members of society who had not undergone specialized training in laboratory science. In the process, proponents of forensic reform asserted much firmer boundaries between those in Chinese society who could claim expertise in forensics and those who could not.

Visibility, Publicity, and Professional Authority

Much as "being public" became a new imperative for many urban associations under the new Republic, as Bryna Goodman has shown, occupational groups that were involved in the investigation of homicide also used newspapers, journals, and other media to articulate their collective identity and social role.[7] A distinction might be drawn at the outset between simple visibility in the media and on the streets and efforts to publicly assert the collective interests and identity of a group. Many of those who were involved in the investigation of crimes were already highly visible in the former sense. Newspapers already covered criminal investigation in detail, and it was not uncommon for police officers, detectives, inspection clerks, procurators, and medical experts to appear as named individuals in this reporting. At the same time, there were very real differences in the ways in which members of these groups used this and other forms of media to engage urban society. Medical journals, for example, provided a powerful resource for those who promoted forensic reform to criticize the judiciary's existing practices and assert the superiority of science-based forensic techniques. Examining the strategies of public engagement that these groups adopted exposes the variable possibilities that existed for these groups to assert their authority in the modern occupational marketplace, both in the context of particular cases and in public discourse more broadly.

Strategies of publicity could take many forms. Take, for example, *Cases of Beiping Detectives* (Beiping zhentan an), a compilation of real cases investigated by members of Beiping's Detective Bureau that was

[7] Bryna Goodman, "Being Public: The Politics of Representation in 1918 Shanghai," *Harvard Journal of Asiatic Studies* 60, no. 1 (2000): 45–88.

published in December 1930 and then again in November 1932 following the popular response to the first printing.[8] The motivation for publishing cases seems to have come from Zhang Ruilin, head of the Second Squadron of the Beiping PSB Detective Bureau, who had broached the project with his long-time acquaintance, senior detective Ma Yulin. Zhang's goal in publishing these cases, as recounted by Ma in his preface to the book, was to "make all circles of society clearly understand how Beiping detectives solve major cases."[9] The detectives of Beiping were already visible in the city's newspapers, which provided detailed coverage of their investigations throughout the 1930s (Figure 4.1). The newspaper *World Daily* even ran a significant multi-part interview with Ma Yulin himself.[10] Publishing cases allowed the squad to go beyond this coverage and actively define its professional identity while responding to the potential criticisms of detractors.

Questions of professional reputation were clearly at issue in the preface written by Xuan Yaojun, an acquaintance of the book's publisher who had been asked to read over and edit the manuscript. While Xuan himself had long enjoyed learning about detection and had delighted in reading detective novels, he had never had a favorable impression of Beijing's detectives. According to Zhu Shixun, who had originally brought the manuscript to Xuan, the negative opinions of detectives held by some in Beijing society stemmed from five sources. These included the ease with which detectives could be falsely slandered, the fact that detectives were often observed in the same seedy environs as criminals, and the perception that detectives used torture to close cases. Negative perceptions also stemmed from detectives' lingering association with the yamen policemen of olden days and the reputation of the latter for implicating innocent people in legal cases. Finally, the bad actions of individual detectives could be taken to represent the entire group, thereby sullying its reputation.[11] The prefaces of the work addressed each of these concerns in detail while showing readers the high value that Beiping detectives placed on "morality" (*daode*) as an important element of their professional ethos. Inspection clerks faced criticisms that were strikingly similar, including their lasting association with the purportedly corrupt yamen

---

[8] The first edition received good publicity, even appearing in an "introducing a new publication" feature in the newspaper *World Daily*. SJRB January 10, 1931, 7. For an overview of Beiping detectives during this period, see Xu, "Wicked Citizens and the Social Origins of China's Modern Authoritarian State," 512–24.

[9] Zhang et al., *Beiping zhentan an, xu* 1. Zhang Ruilin also believed that an edited compilation of cases could serve as a kind of reference work to assist China's detectives, whose knowledge and skill were still undeveloped. Such a work would also assist in correcting the imbalances of China's largely foreign-derived literature on detection (*xushi* 3–4).

[10] SJRB March 12, 1933, 8.   [11] Zhang et al., *Beiping zhentan an, xuwen* 3–6.

4.1 Detectives Ma Yulin and Zhang Ruilin.

underlings of the old imperial state and critics' tendency to take the errors of individuals as representative of the failures of the group as a whole.

Cases of Beiping Detectives also made a strong argument for the special skills and knowledge possessed by detectives. In his preface to the work, Ma Yulin noted that detection had developed into "a kind of specialized science" (*yizhong zhuanmen kexue*) that was always undergoing innovation, a claim that associated detectives with the kinds of authoritative academic knowledge that modern professionals commonly use to legitimize their expertise.[12] At the same time, as Ma reminded readers, detection was a field based on practical experience, not theoretical speculation divorced from actual cases. The cases featured in *Cases of Beiping*

---

[12] Ibid., *xu* 2.

Detectives demonstrated detectives at their finest, using powerful observational skills and connecting disparate clues to solve cases. This was demonstrated, for example, in the book's chapter on the killing of Su Baojie, a case that was taken by some to represent the deficiencies of the procuracy's forensic investigation practices. The treatment of this case, which appeared in *Cases of Beiping Detectives*, focused on the role that detectives played in identifying Su's body after it had been dumped in the western part of the Inner City following the automobile accident that ended his life. The book claimed the closing of this case for the detective squad, describing the process through which a detective noticed that the corpse was missing the very same shoe that had been discovered at the site of the suspicious traffic incident across town. It was on the basis of this observation that detectives realized that the cases were connected and that the dead body was that of Su, who had already been reported missing.[13]

A work such as *Cases of Beiping Detectives* reminds us that the possibilities for "being public" were quite diverse. In this instance, detectives' assertions of collective identity and authority relied on the reading public's considerable interest in detective novels (*zhentan xiaoshuo*), a new genre that the work's preface-writers invoked even as they argued that the cases in the book were more authentic than the entertaining, if idealized, fabrications of Sir Arthur Conan Doyle (1859–1930).[14] This was a very direct engagement with the popular reading preferences of a highly commercialized publishing industry. Other individuals and groups followed strategies of publicity that could include publishing a professional journal, textbook, or case collection demonstrating technical proficiency and expert knowledge. All of these, for example, were employed by the Beijing-based fingerprint expert Xia Quanyin, who in the mid-1920s was involved in the publication of *Fingerprinting Magazine* (Zhiwen zazhi), a primer on Xia's own adaptation of the Henry system of fingerprint classification, and a collection of fingerprint cases that he had personally handled.[15] Mobilized in these ways, the printed word provided detectives and others with a way to address potential criticisms, shape the story of their involvement in particular cases, and claim professional

---

[13] Ibid., 167–74. When compared to the police files and newspaper reports, the book's description contains several inconsistencies and errors. For example, in the book the victim was identified as Su Junchen and "Jin Dasheng," the name of the driver of the vehicle, was written with an alternate spelling.

[14] Ibid., *xuwen* 7, *xu* 2–4, *shushi* 3–4.

[15] Xia Quanyin, *Zhiwen xueshu* [The academic learning of fingerprinting], in *Zhentan congshu* [Collectanea on detection], ed. Xia Quanyin et al. (Nanjing: Jinghua yinshuguan, 1935); Xia Quanyin, *Zhiwen shiyan lu* [A record of practical demonstrations of fingerprinting] (Beijing: Zhongguo yinshuju, 1926).

legitimacy, concerns that were not insignificant in the highly visible and at times precarious task of homicide investigation. By implication, groups that did not attain the same possibilities for self-representation could easily face the challenges of defending themselves and their reputations.

Inspection Clerks under Fire

Within the reams of newspaper reporting on crime that circulated in 1920s China, the forensic practices of the judiciary garnered a great deal of attention, as did the officials and inspection clerks who were involved in particular cases. Those cases in which errors had been made during the forensic examination, or in which the evidence simply seemed doubtful, became public targets for individuals and groups with an interest in the state of China's forensic investigation practices. If during the late imperial period it had primarily been the official bureaucracy that supervised forensic examinations and evaluated the integrity of the proceedings, by the second decade of the Republic a range of other groups were also weighing in on the state's forensic practices, deliberating on whether errors had been committed in particular cases and whether there were better techniques or sources of expertise that should have been used. Of course, disputes over particular forensic findings or techniques were hardly new. As we saw in Chapter 3, the Republican judiciary routinely faced the prospect that relatives of the deceased or others might disagree with forensic findings and even put forward alternate interpretations of the forensic evidence. Yet, the criticisms that emerged over the 1920s invested forensic investigation with new social and political meanings that reflected the broader moment of nationalism, profession-building, and social reform that defined this period in Beijing and other Chinese cities. In this intellectual and social milieu, errors real or perceived could easily be taken as evidence not simply of the malfeasance of individual examiners but also of the backwardness of China's judicial practices and the country's failure to adopt new forms of scientific knowledge and expertise. These critical discourses challenged the legitimacy of inspection clerks by raising questions about their suitability for the new, modern challenges that China faced.

One example of a highly visible case that raised questions about the judiciary's forensic practices was that of Liu Lianbin, a young woman who was found dead in July 1923 at the silkworm egg farm at which she had been working in Wuxi county, in southern Jiangsu.[16] The silkworm

---

[16] For an edited collection of documents from the case, see "Jiangsu Wuxi Liuan zhi huizhi" [Collected records from the case of Liu Lianbin from Wuxi, Jiangsu], *Minguo yixue zazhi* 2,

breedery was run by a formally trained agricultural expert and former faculty member of Southeastern University named He Kang, who took Liu on as a trainee after two Southeastern students, who were from Liu's native Sichuan, put her in touch with him. The initial examination of Liu's body, conducted by a trial officer and inspection clerk from the Wuxi county government office, found that Liu had hung herself. This verdict was immediately challenged by Liu's brother, who suspected that the manager of the breedery had sexually assaulted and possibly murdered her. The body was exhumed and a second examination was carried out under the auspices of the Wu County Local Procuracy, one of its forensic examiners, and a foreign physician who was reportedly ill-equipped to conduct such an examination.[17] In the end, Wuxi authorities ruled that it was a case of suicide, even though they did find He Kang guilty of illegally taking possession of letters and money sent to Liu by her acquaintances at Southeastern University.

From the beginning of the case, numerous individuals from Liu Lianbin's native Sichuan pushed judicial authorities to investigate the possibility that she had been murdered. Liu's supporters attended the second examination following the exhumation, taking detailed notes that raised questions about the integrity of the findings.[18] Almost a year after Liu Lianbin's death, a "Committee for Redressing Injustice in the Case of Liu Lianbin" was still attempting to drum up support following the Liu family's unsuccessful attempt to appeal the initial verdict.[19] Throughout the handling of this case, Liu's supporters publicly challenged the authority of legal officials and their forensic personnel by casting doubt on their interpretations of the body evidence. Points of contention included wound-like marks that were observed on Liu's badly decomposed body as well as the effluvia, including blood and other matter from decomposing internal organs, that had been discovered during examinations of Liu's genitalia. Both, Liu's supporters argued, suggested the possibility that she had been sexually assaulted before her death. Liu's supporters also criticized judicial officials' use of the *Washing Away of Wrongs*, which they portrayed as outdated and lacking a scientific basis.[20]

In such cases, the perceived injustices of individuals became linked to much larger questions of judicial reform, a national issue heralded by professional associations and other groups and that appeared often in the

---

no.1 (1924): 46–54. The verdict rendered by the magistrate and trial officer of Wuxi county in November 1923 was printed in *Falü zhoukan*, nos. 22–5 (1923).

[17] This was the impression given in the account of one of Liu's advocates who was present at the scene. "Jiangsu Wuxi Liuan zhi huizhi," 48.

[18] Ibid., 48–9.   [19] CB May 12, 1924, 6.   [20] "Jiangsu Wuxi Liuan zhi huizhi," 49–50.

press during the 1920s. The Washington Conference (November 1921–February 1922), which addressed the China-related interests of the United States, Britain, and Japan in the changed geopolitical landscape that followed WWI, had mandated a commission tasked with inspecting the condition of China's legal system as a precondition for modifying the countries' extraterritorial judicial claims.[21] This impending investigation, which, after delays, finally took place over a number of months in 1926, deeply informed the public discourse on forensics during the early–mid 1920s. For those who viewed cases such as that of Liu Lianbin as demonstrating China's failure to adopt what were understood to be the most modern standards of forensic expertise, it was easy to make the case that this area of judicial reform had severely lagged and that the use of inspection clerks and their old practices diminished the integrity of the law at exactly the moment when demonstrating progress was essential. In pressing for continuing investigation in the case of Liu Lianbin, for example, supporters made their own call to arms in a statement that appeared in *Morning Post*, asking "at the moment when foreigners are paying attention to the administration of justice in China and our compatriots are summoning each other to action to withdraw consular jurisdiction, how can all of the conscientious persons of the nation not take an interest when the iniquity of the dignified high judicial authorities of Jiangsu is as egregious as this?"[22]

In this context, while inspection clerks were members of a group that was already publicly visible in Beijing, as we have seen, they did not usually promote their collective occupational interests through the written word. We find one extraordinary exception in a document penned by the senior forensic examiner Yu Yuan, who was responding at the time to proposals made by the Beiping Bar Association to reform the city procuracy's forensic practices.[23] The Bar Association's proposals, sent in late July 1928, admonished Beijing judicial authorities to ensure that forensic personnel were adequately trained and urged the procuracy to seek the assistance of those with medical expertise in particularly difficult

---

[21] *Report of the Commission on Extraterritoriality in China* (Washington: Government Printing Office, 1926).

[22] CB May 12, 1924, 6. For another call for forensic reform that invoked the impending investigation, see Jiang Zhenxun, "Diaocha sifa shengzhong ying zhuyi fayi zhi wujian" [My opinion that legal medicine should be paid attention to amidst the investigation of the judiciary], *Minguo yixue zazhi* 4, no. 2 (1926): 43–6.

[23] Beiping Bar Association to Beiping Local Court and Procuracy, July 31, 1928, BMA J174-2-152, 43–54; Statement of Yu Yuan, August 3, 1928, BMA 174-2-152, 56–8. For an analysis of this same document that focuses on Yu Yuan's use of the concept *jingyan* (experience) as a point of interaction between older conceptions of empirical knowledge in late imperial forensics and new discourses of science, see Asen, "The Only Options?"

cases. In explaining its grounds for making these suggestions, the Bar Association questioned not only inspection clerks' knowledge in forensics, but also their "moral character" (*daode renge*), a charge that resonated with long-standing discourses on the lack of moral integrity of the *wuzuo* of late imperial times. As evidence of the dangers of entrusting legal cases to such people, the Bar Association cited the case of Su Baojie, bringing up the fact that the individual who conducted the initial forensic examination in the case had claimed that there was a sexual angle to the killing. The Bar Association's implication was that the true nature of the case would have been revealed more quickly if the procuracy had entrusted Su's body to a physician or other forensic specialist from the beginning.

In his response to the Bar Association's letter, Yu Yuan addressed the organization's criticisms directly. Underlying Yu's response was an assertion that inspection clerks deserved the same treatment as other occupational groups: the mistakes of individuals should be treated as such, not as an indictment of the group as a whole and certainly not an indictment of the group's "moral character." Discussing the case of Su Baojie, Yu acknowledged that the initial examination of the body had been flawed, but noted that the inspection clerk who made the error was inexperienced and had been subsequently suspended. Yet, while Yu conceded that one could say that this individual was unskilled, it was unreasonable to impugn his moral character. The larger implication that can be detected in Yu's response was a criticism of the prevailing tendency to interpret the actions of individual inspection clerks as demonstrating the failings of the entire group. Later in the letter Yu even asked whether any members of the Bar Association had ever made errors in the course of their own professional duties or had ever acted against the law. If so, Yu asked, "is it the case that it is the individual's moral character alone that is base or that of the entire association?" The Bar Association had thus failed to distinguish between malfeasance committed intentionally and that committed in error while mistakenly applying the failings of an individual to the group as a whole.

Inspection clerks were hardly alone in engaging in occupational error or malfeasance, whether real or perceived. The difference was that other groups were adopting strategies of publicity to address the concerns, to explain away the controversies, and to argue for their professional legitimacy. This was the point, for example, of a piece written by Sun Kuifang, a French-trained expert in legal medicine who was tapped by the Nationalist party-state to plan the central government's Research Institute of Legal Medicine, an institution that would become the foundation for China's legal medicine profession during the 1930s.

At a moment when preparations for this institution were getting under-way, Sun became implicated in the controversial case of a young woman named Xuan Axiang, who died suddenly following an argument with her husband. In carrying out the initial examination of Xuan's body for Shanghai authorities, Sun found that she had suffocated, likely from being strangled by her husband. When a new witness statement and chemical testing results indicated that Xuan had died from accidental suffocation following loss of consciousness from having taken opium, Sun was put in the difficult position of reversing his earlier autopsy findings. Sun's lengthy defense of his role in this case was published in both a medical journal and the major newspaper *Shenbao*.[24] By explaining that these examinations had both been consistent with the facts as they were known at the time, Sun hoped to correct the public's assumptions that his examinations had been inconsistent or even faulty. Sun saw much at stake in correcting such misperceptions, noting that "if society loses trust in legal medicine from the outset, there will be major consequences for the prospects of judicial reform."

In this context, it is noteworthy that Yu Yuan's defense of the city's forensic personnel was not publicly disseminated in newspapers or other printed texts. Explicit defenses of inspection clerks and the *Washing Away of Wrongs* rarely were (a stunning exception is discussed in Chapter 6). Inspection clerks were hardly unique in being branded, at times very publicly, as holdovers of the old imperial age. Beijing's yinyang masters, for example, became a focus of hygiene reformers' criticisms for impeding the establishment of a medicalized death registration system and even-tually lost their long-standing authority to certify deaths for the police.[25] The case of physicians of Chinese medicine, who were likewise criticized for obstructing the spread of modern medical customs in China, suggests a completely different outcome. After facing a major attack in 1929 – led by Yu Yunxiu (1879–1954), a Japan-trained physician and major propo-nent of Western medicine in China, and backed by the advisory board of the Nationalist government's newly established Ministry of Health – physicians of Chinese medicine effectively mobilized popular support behind their cause, in part by appealing to nationalism.[26] Inspection clerks lacked the resources, sources of collective occupational identity, compelling claims to nationalism, and political connections of this move-ment. At the same time, the increasingly visible profile of advocates of forensic reform, many of whom were physicians of Western medicine or

[24] "Xuan Axiang an jianyan jingguo" [Circumstances of the forensic examination in the case of Xuan Axiang], *Yiyao pinglun*, no. 45 (1930): 24–9; SNB November 10, 1930, 10.
[25] Yang, *Zaizao "bingren,"* 164.
[26] Xu, *Chinese Professionals and the Republican State*, 190–214.

even experts in legal medicine such as Sun Kuifang, further associated the world of judiciary-centered forensics with a sense of backwardness, in the process challenging its legitimacy.

Defining a Discipline: The Professional Identity of Legal Medicine

During the 1920s, physicians of Western medicine, more than the members of any other professional group or association, transformed the debate over China's forensic practices, arguing forcefully that the judiciary's existing practices had to be replaced with those modeled on the forensic institutions of continental Europe and Japan, which depended heavily on the involvement of experts in legal medicine and other fields of modern Western science. The case of Liu Lianbin, for example, became a focal point for medical reformers' discussions of forensics on the pages of *Republican Medical Journal* (Minguo yixue zazhi), a journal that reflected the interests of the significant community of Chinese physicians of Western medicine who had been trained in Japan. The journal published a collection of documents pertaining to the forensic evidence in the case, as well as essays criticizing officials' handling of the case and advocating the reform of China's forensic practices.[27] For example, a piece titled "Why not establish forensic medicine experts?" appeared in the pages of this journal and in the important literary supplement of *Morning Post*.[28] The author, Gong Shucang, targeted the Beijing procuracy specifically, describing its continuing reliance on inspection clerks as a refutation of those modern forensic practices, rooted in Western science, that were followed in other countries. Gong cited the cases of Liu Lianbin as well as the January 1924 murder of Liu Ma, examined in Chapter 5, as evidence of the judicial injustices that could result from this state of affairs. These were cases, Gong wrote, in which "the corpse was examined repeatedly with delays going on for months and no conclusive resolution in the end. Because of this, many innocent people were implicated and left groaning for a long time in the dark chambers of prison." Like other proponents of forensic reform who cited botched and uncertain cases as evidence of the fundamental failings of China's forensic practices, Gong failed to acknowledge the complexity of the investigation

---

[27] "Jiangsu Wuxi Liuan zhi huizhi"; Gong Shucang, "Weishenme bu she caipanyi?" [Why not establish forensic medicine experts?], *Minguo yixue zazhi* 3, no. 5 (1925): 229–30; Hou Yuwen, "Zhi Sifa bu zhi chengwen" [A petition sent to the Ministry of Justice], *Minguo yixue zazhi* 4, no. 1 (1926): 2–3; Jiang, "Diaocha sifa shengzhong ying zhuyi fayi zhi wujian."

[28] Gong, "Weishenme bu she caipanyi?"; *Chenbao fukan* March 29, 1925, 3–4.

process that led to these outcomes. In the killing of Liu Ma, for example, the examination of the body was in fact the least contested piece of evidence in a case that included a confusing crime scene covered in fingerprints and blood stains that were interpreted and reinterpreted throughout the course of the case.

One of the most important advocates of legal medicine during this period was Lin Ji, a graduate of the National Medical College in Beijing (later reorganized as the medical school of Beiping University), and subsequently a faculty member of this same institution who was called upon to examine Zhang Shulin's dismembered remains in early 1936. Lin Ji was very much a product of this school, its traditions of Japanese–German medical training, and the emphasis on public engagement with state and society that it fostered among its faculty and students.[29] Following completion of his medical studies in 1922, Lin Ji remained at the school as an assistant in pathology in a department that was developing a reputation for providing appraisals of forensic evidence in legal cases.[30] The school soon sent Lin Ji to Würzburg University in Germany, where he received his M.D. while pursuing advanced studies in legal medicine. During this period Lin Ji developed an interest in forensic applications of entomology, among other subjects, and wrote articles that introduced these areas of research to readers in China.[31] Returning to China in summer 1928, Lin Ji went on to pursue a series of profession- and state-building activities under the auspices of the newly established Nationalist state in Nanjing. It was at this point that Lin Ji took over the planning of the central government's Research Institute of Legal Medicine from Sun Kuifang and brought it to completion in 1932.

Throughout the course of his career, Lin Ji put a great deal of effort into publicly explaining the importance of legal medicine and the varied contributions that this scientific discipline could make to the modern state. The writings that Lin published in professional journals and other venues for educated readers often contain basic explanations of the nature of a discipline that, as Lin himself recognized, was not well known in

[29] For more on medical instruction at this school, see Daniel Asen and David Luesink, "Globalizing Biomedicine through Sino-Japanese Networks: The Case of National Medical College, Beijing, 1912–1937," in *China and the Globalization of Biomedicine*, ed. David Luesink, William H. Schneider, and Zhang Daqing (under review). For a biography of Lin Ji, see Huang Ruiting, *Fayi qingtian: Lin Ji fayi shengya lu* [A righteous medico-legal expert: A record of the career of the medico-legal expert Lin Ji] (Beijing: Shijie tushu chuban gongsi, 1995).

[30] List of instructional staff, 1922, BMA J29-1-9, 13.

[31] Lin Ji, "Zuijin fayixuejie jiandingfa zhi jinbu" [Recent progress in medico-legal appraisal methods], *Zhonghua yixue zazhi* 12, no. 3 (1926): 220–37; Lin Ji, "Fayixue sizhong xiaoshiyan" [Four kinds of small experiments in legal medicine], *Guoli Beiping daxue yixue niankan* 1, no. 1 (1932): 297–315.

China.[32] Lin Ji commonly wrote that legal medicine could serve all areas of modern governance, whether legislation, the administration of justice, or governmental administration and policing. Legal medicine could provide assistance in everything from the drafting of medical laws to forensic investigation to a range of other applications in administering the population's health and responding to crime, disaster, and other social pathologies. The forensic examination of evidence was, in this sense, a relatively narrow application of a field that could serve as the foundation for an expansive form of physician-led scientific governance. Lin did not claim public hygiene as falling within the broad mandate of legal medicine, even though China's first medico-legal laboratories, as in the case of some institutes of legal medicine in Europe, offered services that would contribute to the surveillance and control of infectious disease.

As explained by Lin Ji and others in medical journals and newspapers, the professional authority of legal medicine rested on its claims of superior knowledge of physical things, including bodies – an "elitist epistemology," in the words of John Harley Warner, that lay at the heart of their arguments for replacing inspection clerks with trained specialists in this field of medicine.[33] The notion that legal medicine took Western scientific medicine and the natural sciences as its "foundation" (*jichu*), as Lin Ji was fond of explaining, thus signaled the basic source of the discipline's epistemological authority.[34] This meant that, in practice, the only valid understandings of bodies and things were those that were accepted by fields of scientific knowledge such as pathology and analytical chemistry, both of which were foundational to medico-legal experts' engagements with physical evidence. Thus, claims about cause of death that had not been proven by an anatomical examination of the internal organs or claims about the use of poison that had not been verified by means of chemical tests were viewed, according to these epistemological standards, as being less certain than those that relied on these fields of modern science. By implication, there would be much firmer boundaries between the experts – that is, professional scientists with formal training and

---

[32] Lin Ji, "Sifa gailiang yu fayixue zhi guanxi" [On the relationship between judicial reform and legal medicine], *Chenbao liu zhou jinian zengkan*, 5th edition (January 30, 1925): 48–53; Lin, "Zuijin fayixuejie jiandingfa zhi jinbu."

[33] The phrase comes from Warner's study of the changing conceptions of epistemological authority in American medicine that accompanied the late nineteenth-century shift from popular empiricism to a new notion of exclusive laboratory-based knowledge. John Harley Warner, "The Fall and Rise of Professional Mystery: Epistemology, Authority, and the Emergence of Laboratory Medicine in Nineteenth-Century America," in *The Laboratory Revolution in Medicine*, ed. Andrew Cunningham and Perry Williams (Cambridge: Cambridge University Press, 1992), 112, 140–1.

[34] Lin, "Zuijin fayixuejie jiandingfa zhi jinbu," 220.

experience in the laboratory – and those who lacked these skills and knowledge.

The idea that the evidence underlying legal judgments should be guaranteed by the expertise of trained professionals and grounded in the epistemological authority of the sciences was a compelling one in 1920s urban China. Indeed, by the mid-1920s it was becoming less and less unusual for Republican courts to seek the assistance of physicians or other outside experts in criminal investigations and trials. The National Medical College in Beijing, for example, had been assisting judicial authorities in cases of suspected poisoning and examinations of blood evidence since the late 1910s and continued to do so throughout the 1920s and 1930s.[35] Over a year beginning in summer 1924, judicial authorities in Shanghai entered into an agreement with Tongji medical school to provide autopsies in their homicide investigations, an arrangement that regularized Shanghai procurators' access to this forensic technique. Beginning in the early 1930s, as director of the Research Institute of Legal Medicine and legal medicine department of the medical school of Beiping University, Lin Ji played an important role in facilitating law courts' access to these services, thereby expanding the Republican judiciary's use of scientific evidence.

While these examples, discussed in later chapters, might suggest that the program of forensic reform that Lin Ji and others were pushing forward was ultimately successful, in practice it was a project that was beset by ambiguities. The forensic authority of Western-medicine physicians and experts in legal medicine rested on the claim that hard boundaries existed between modern Western science (*kexue*) and other forms of knowledge that did not meet the rigorous standards of experimentation and theorization that were associated with this new epistemological category. Proponents of forensic reform thus argued that the *Washing Away of Wrongs* contained inaccurate knowledge and misguided forensic practices and, for these reasons, that judicial officials and inspection clerks should give up their forensic responsibilities in favor of a system of forensic investigation staffed by physicians. Yet, under circumstances in which there was a paucity of medical experts and in which the judiciary was equipped to make its own claims over forensic investigation, assumptions about medical experts' absolute and exclusive authority tended to yield to more amorphous forms of inter-professional cooperation in which experts such as Lin Ji worked alongside the judiciary's own forensic personnel, at

---

[35] Tang Erhe, "Xue fazheng de ren keyi budong xie yixue ma?" [Can those who study law and politics not also understand a bit about medicine?], *Xin jiaoyu* 2, no. 3 (1919): 295–303.

times even examining evidence in the same cases. This kind of forensic pluralism generally defined the working arrangements that developed between judicial officials and medico-legal experts in 1920s and 1930s China. In this context, medico-legal experts struggled to make judicial officials' own forensic practices more compatible with their own, a necessary precondition for exerting greater influence over the handling of forensic evidence within the judiciary. Educating judicial officials and others about legal medicine and the discipline's new norms of evidence collection and analysis thus became an important aspect of promoting forensic reform and claiming professional authority.

Popularizing Legal Medicine

In late May 1934, Xu Songming (1890–1991) gave an address at the Research Institute of Legal Medicine before a pilot class of medical school graduates who were receiving specialized training at this new institution.[36] Following its establishment in August 1932, the Research Institute had quickly become an important institutional force for the development of legal medicine in China. Under the directorship of Lin Ji, this facility supported a range of activities including investigating forensic cases in Shanghai and local courts across China, providing advanced training in legal medicine, and promoting the authority of medico-legal experts in the judiciary. The speech was meant to encourage these future forensic specialists while addressing the challenges that were facing a profession still in its infancy. Having received training at the elite Kyūshū Imperial University College of Medicine, Xu had been one of the Japan-trained physicians who comprised the early cohort of teaching faculty at National Medical College in Beijing in the late 1910s and 1920s. An expert in pathology and legal medicine, Xu had provided appraisals of physical evidence to judicial authorities in Beijing and elsewhere, even playing a key role in investigating the murder of Liu Ma in the mid-1920s, as we will see in Chapter 5. At the time of this speech, Xu was serving as president of Beiping University, one of many administrative positions that he would hold over the course of his career. The text of the speech was published in the *Monthly Bulletin of Legal Medicine* (Fayi yuekan), the de facto mouthpiece of legal medicine professionals during the 1930s.[37] The journal featured research articles written by Research Institute personnel, didactic pieces meant to introduce

---

[36] "Zenyang zuo fayishi ji fayi zai Zhongguo zhi chulu" [How to be a medico-legal physician and the prospects for legal medicine practitioners in China], *Fayi yuekan*, no. 6 (1934): 1–4.

[37] In April 1936 the Research Institute's journal began to be published as a quarterly publication titled *Quarterly Bulletin of Legal Medicine* (Fayixue jikan).

procedures of medico-legal investigation, and examination reports from actual cases, some of which documented the purportedly unscientific examination practices of China's legal officials and inspection clerks. Publishing such a journal made a powerful statement about the professional identity and public role of this new community of experts. In this sense, it represented a maturation of the push for forensic reform that had begun about a decade earlier.

Xu began by recounting a case that he had handled while at National Medical College in Beijing. The case had involved the murder of the wife of a traveling businessman in Shanxi province. Local people suspected the husband of committing the murder, and investigating authorities had found a piece of clothing covered in small brown marks that, it seemed, could have been the victim's blood. Shanxi officials sought assistance from the National Medical College and asked experts at this institution to determine whether or not the marks were blood stains. For Xu, the way in which the officials had submitted this request indicated their limited understanding of the science of blood analysis and the capabilities of experts in legal medicine. In making their request, they had simply asked whether or not the substance was blood, failing to ask Xu to confirm that the blood was from a human and not animal blood, one of the many specific questions that were fundamental to the investigation of blood evidence in legal medicine. While undoubtedly a minor oversight on the part of the officials, the fact that they had failed to be more precise in their request symbolized for Xu the low status of legal medicine in Chinese courts and the general failure of judicial authorities to accept the evidentiary norms of medical experts in legal cases. Such officials lacked the basic understanding of legal medicine that would allow them to know the right questions to ask of their evidence when seeking the assistance of an expert.

This failure of cooperation, Xu argued, was indicative of the challenges facing legal medicine in China at that moment. These included the courts' failure to acknowledge the importance of legal medicine and the resulting low status of the profession in the eyes of China's jurists. That the profession's members had failed to differentiate themselves from the judiciary's forensic personnel further gave it a low status in the eyes of many, including medical students who did not want to pursue such a career. The problem would be remedied when judicial authorities finally recognized the crucial status and full capabilities of this profession. This low status would also be remedied, Xu noted, when ordinary people developed trust in legal medicine. Thus, Xu urged the trainees of the Research Institute to work diligently and "handle some amazing cases for society, making ordinary people understand that the new legal medicine experts can really handle cases that could never be solved before."[38]

Building legal medicine thus involved more than simply explaining its importance to the readers of medical journals who likely already supported the expanded societal influence of Western scientific medicine in general. Most immediately, it required establishing a niche for itself among the judicial officials who might use forensic testing services in their cases and legitimating itself within Republican society more broadly. This involved cultivating broad public understanding of the profession as well as of the conceptions of scientific knowledge on which its claims of exclusive authority were based. Creating a demand for medico-legal services among judicial officials also required that they have some understanding of the technical knowledge that informed the forensic work of the experts – the very knowledge that, by virtue of its esoteric and specialized nature, justified the involvement of trained professionals in the first place. This notion of laypersons' elementary understanding of the knowledge of specialized fields such as legal medicine was usually referred to as "common knowledge" (*changshi*), a concept that was often used when describing popularized understandings of hygiene, medicine, law, and other areas of modern knowledge.[39] It had been precisely their lack of such knowledge, Xu argued in his speech, that had led the Shanxi authorities to ask imperfect questions of their evidence. An implicit goal of disseminating basic knowledge of legal medicine was thus to facilitate inter-professional coordination between legal officials and medico-legal experts by having the former adopt the norms of evidence collection and analysis of the latter.

Such "common knowledge" could be disseminated in different ways. Lin Ji himself authored didactic texts meant to instruct legal authorities who were tasked with criminal investigation and adjudication and that could be used in training classes for these officials.[40] Legal officials also received this knowledge through the written reports that experts such as

---

[38] "Zenyang zuo fayishi ji fayi zai Zhongguo zhi chulu," 3.

[39] For example, disseminating "common knowledge" of medicine and hygiene was the goal of various initiatives carried out by faculty and students of the National Medical College in Beijing, including public exhibitions of anatomical specimens; i.e., SJRB January 25, 1932, 7; Asen and Luesink, "Globalizing Biomedicine through Sino-Japanese Networks." For the dissemination of basic legal knowledge, see Jennifer Altehenger, "Simplified Legal Knowledge in the Early PRC: Explaining and Publishing the Marriage Law," in *Chinese Law: Knowledge, Practice and Transformation, 1530s to 1950s*, ed. Li Chen and Madeleine Zelin (Leiden: Brill, 2015).

[40] See, for example, Lin Ji's published teaching materials for the Institute for the Training of Judicial Officials (Faguan xunliansuo) of the Ministry of Judicial Administration. Lin Ji, *Fayixue gelun* [A detailed discussion of legal medicine] (Nanjing: Sifa xingzheng bu, 1930).

Lin Ji returned to them after examining their evidence in the laboratory. These documents, which could include an explanation of the relevant scientific principles in addition to discussion of the evidence in the case, were an important mode of communication between forensic laboratories and judicial officials. The Research Institute also attempted to change legal officials' evidence collection practices by having the Ministry of Judicial Administration, the Nationalist government's central judicial agency, endorse laboratory-based examination techniques through official orders.[41] In all of these efforts there was a close relationship between educating officials about legal medicine and claiming exclusive authority for members of the discipline – a dynamic that seems natural, perhaps, given medico-legal specialists' advanced training in a body of academic knowledge that was not well known to Chinese officials or society. At the same time, as Jennifer Mnookin has argued in her study of handwriting experts in the United States, "educating" the law's fact-finders can constitute a powerful strategy for making them see evidence in new terms and, ultimately, for claiming professional authority.[42] Indeed, the point of disseminating basic knowledge of legal medicine to judicial officials was not to encourage them to use this knowledge to solidify their own position in China's forensic examination system, but, rather, to facilitate their coordination with the "real" experts, thus consolidating professional scientists' exclusive authority over this area of judicial practice.[43]

During the 1930s, one of the accomplishments of the Research Institute of Legal Medicine as well as the legal medicine department of the medical school of Beiping University (where Lin Ji served as director from 1935–7) was to expand officials' access to laboratory testing services, a development examined in Chapter 7. These new connections between the medical and legal professions were made publicly visible in the print media, which picked up on cases handled by these new facilities as well as general news of the field. Readers of *Shenbao*, for example, might have learned of the new forensic testing services provided by the

[41] For example, see an order issued by the Ministry of Judicial Administration that all high courts and procuracies direct lower courts to give up the older methods for examining suspected blood stains contained in the *Washing Away of Wrongs*. Hebei High Court to Beiping Local Court, October 1936, BMA J65-3-300, 16–19.

[42] Jennifer L. Mnookin, "Scripting Expertise: The History of Handwriting Identification Evidence and the Judicial Construction of Reliability," *Virginia Law Review* 87, No. 8, Symposium: New Perspectives on Evidence (2001): 1723–845.

[43] For more on the broader implications of popular engagements with science for questions of professional authority, ideologies of modernity, and state power in twentieth-century China, see Schmalzer, *The People's Peking Man*; Fa-ti Fan, "Science, State, and Citizens: Notes from Another Shore," *Osiris* 27, no.1, Clio Meets Science: The Challenges of History (2012): 227–49.

Research Institute as well as the graduation of the class of forensic specialists that Xu Songming had addressed in his speech.[44] Much like the messages that Lin Ji and Sun Kuifang were disseminating through professional journals and other media, newspaper reporting too might portray scientific knowledge as a necessary foundation for the most authoritative kinds of forensic expertise while questioning the value of techniques that did not meet this new epistemological standard. For example, in mid-April 1933, *Shenbao* published Lin Ji's criticisms of the use of a silver hairpin or needle to confirm death by poisoning, a technique of the *Washing Away of Wrongs* that would become a target of sustained criticism as forensic laboratories in Beiping and Shanghai expanded their influence over poisoning cases during the Nanjing decade.[45] The piece recapitulated Lin Ji's argument that the technique could not serve as a reliable test for poison and that only those with specialized expertise in chemistry could hope to make accurate claims about the evidence in such cases. This was a more "popular" version of the same arguments that were being made on the pages of the Research Institute's own *Monthly Bulletin of Legal Medicine*, as well as in other medical journals.

It is difficult to gauge the extent to which the new visibility of legal medicine in *Shenbao* and other newspapers actually changed the forensic norms and expectations held by readers of the urban press. This reporting was counteracted, to some extent, by the prevalence of coverage of the judiciary's own forensic practices in the same media. During a period in which inspection clerks continued to examine the body in most cases, and in which judicial officials continued to accept the *Washing Away of Wrongs* as a valid source of evidence, it was not uncommon for newspaper coverage of forensics to focus largely on the body examination practices of judicial officials, not the less widespread forensic work of China's small community of legal medicine experts. This coverage legitimized the role that inspection clerks played in these cases while implicitly contradicting the claims of medico-legal reformers that forensic examination practices associated with the *Washing Away of Wrongs* lacked validity and effectiveness. Indeed, in much reporting there was nothing to indicate that these older techniques did not still work or that they should not be used because of their lack of grounding in scientific knowledge. In *Morning Post*, for example, it was not uncommon for reporters to simply reproduce the sequence and terminology that inspection clerks used to describe the body in their cases, a journalistic practice that transmitted to readers a conception of the body quite different from that which physicians of

---

[44] SNB August 2, 1934, 11; SNB December 21, 1934, 11.   [45] SNB April 19, 1933, 10.

Western medicine were promoting.[46] Even the technical concepts of "vital spots" (*zhiming zhi chu*) and "mortal wounds" (*zhiming shang*), basic elements of the judiciary's official examination regime, appeared in this reporting, not as old techniques to be questioned under the critical gaze of science, but as unremarked-upon elements of the "facts" of legal cases, a status that they enjoyed in court as well.[47] Thus, as much as newspapers played an important role in the attempts of new groups of professionals to assert their authority, their coverage also tended to accurately depict the pluralism that defined China's forensic scene in the 1920s and 1930s, itself a reflection of the fact that the judiciary continued to invest authority in forensic examination practices that were not based on Western scientific medicine.

Conclusion

We might conclude this chapter by comparing two photographs of forensic practice. The first, taken within the walls of the medical school of Beiping University, depicts Lin Ji examining skeletal remains in a space that is identified as an autopsy room (Figure 4.2). Lin Ji, clad in a protective laboratory garment, pensively studies a piece of bone. A skull is sitting on the examination table in front of him, and other apparatus is visible behind him. Stands in the background of the image seem to be benches for observers, possibly medical students. The photograph was one of several published in summer 1936 in three special issues of *New Medicine* (Xin yiyao zazhi), a medical journal published by the Medical and Pharmaceutical Association of the Chinese Republic (Zhonghua minguo yiyaoxue hui) (Figure 4.3).[48] Aside from these photographs and an introductory essay on legal medicine penned by Lin Ji himself, these issues reprinted fifty of the reports that Lin Ji returned to officials in north China after examining their evidence in cases. This publication comprised part of a very public argument for the necessity of laboratory-based forensic expertise – a campaign that included direct pleas to local officials and central judicial organs, as well as publication of cases and other material in medical journals. According to Lin Ji and, by the mid-1930s, a growing

---

[46] CB May 7, 1924, 6; CB May 10, 1926, 6.

[47] For example, see a *Shenbao* article that describes a victim's wounds as "mortal" (*zhiming*) and "non-mortal" (*fei zhiming*), technical terms in the judiciary's official body examination routine. SNB December 16, 1928, 16. The piece was titled "A thorough forensic examination conducted in a case involving a murder for money, a mortal wound on the *taiyangxue*" (Moucai haiming an zhi yanming, zhiming shang zai taiyangxue). *Taiyangxue* denoted a specific location on the temple as designated in the judiciary's official examination forms.

[48] *Xin yiyao zazhi* [New Medicine] 4, nos. 5–7 (1936).

剖　　驗　　室
（ 骨　髏　檢　查 ）

4.2 Autopsy room, Institute of Legal Medicine, Beiping University Medical School.

community of like-minded physicians, Chinese law could not do without medico-legal experts or the laboratory facilities in which they worked.

The second photograph is an almost incidental representation of a forensic examination included in *Weekly Pictorial Supplement of Capital News* (Jingbao tuhua zhoukan) in early February 1933 (Figure 4.4).[49] Inset on a page with other news stories, the image is of an unidentified man bending over the body of Wang Weisan, a journalist for the Shanghai newspaper *China Times* (Shishi xinbao) who had been assassinated in Nanjing on January 31 in a notorious and unresolved murder that left

[49] *Jingbao tuhua zhoukan* [Weekly Pictorial Supplement of Capital News], issue 184, February 12, 1933, 3.

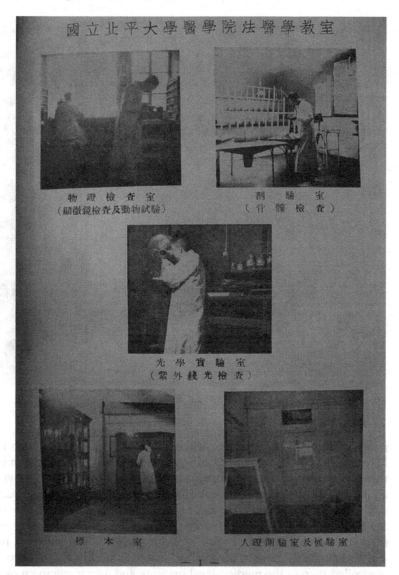

4.3 Photographs of Institute of Legal Medicine, Beiping University Medical School.

4.4 Forensic examination of Wang Weisan's body.

journalists and others calling for officials to investigate and apprehend the killers.[50] The caption reveals that the photograph was taken at the time of the forensic inspection of Wang's body. While no information is provided about the person who appears to be looking intently at a particular part of the middle or lower body, the large sheet of paper in his hand (possibly a copy of the official body examination form) suggests that he was an inspection clerk. The scene of this examination was very different from Lin Ji's laboratory space in the medical school. The environment was less controlled than in the medico-legal institute and, occurring in an ostensibly public place, spectators watched in the background. Unlike the pensive Lin Ji, ensconced in the enclosed space of the autopsy room,

---

[50] The killing and its aftermath received intensive press coverage; i.e., SJRB February 2, 1933, 4; SJRB February 4, 1933, 4; SJRB February 6, 1933, 5; SNB February 2, 1933, 11.

the examiner in this image seems to be conducting the inspection of the body outside.

These photographs suggest some of the differences in the ways in which those who investigated homicide in Republican China engaged new urban publics, as well as the import of such engagements for questions of professional authority. The photograph of Lin Ji was included in a specialized medical publication meant to promote the cause of Western scientific medicine and legal medicine. It was part of an argument, made to judicial officials and others, that China needed more examiners like Lin Ji. Similar arguments were made in other printed texts, including work reports demonstrating the successes of legal medicine testing facilities and the journal *Monthly Bulletin of Legal Medicine*, which published cases and other profession-related materials. The photograph of Wang Weisan's body examination was different, reproduced as one photograph among numerous others and meant to sell newspapers. While inspection clerks became visible, almost despite themselves, in the considerable output of industrial print capitalism in China, their public visibility was a factor of commercial publications' interest in crime and the everyday happenings of urban society. It was not the result of a public campaign to promote their collective professional interests, but, rather, reflected the ease with which corpse examinations could become the focus of sensational news coverage. This coverage had the unintended effect of disseminating knowledge of the judiciary's forensic examination techniques and legitimizing procurators and inspection clerks' involvement in homicide cases.

This chapter has examined the ways in which the public circulation of information about homicide investigation became an arena for new professional politics during the 1920s and 1930s. Turning now to a notorious murder that occurred in Beijing in 1924, we will examine the ways in which these same groups negotiated their claims over the investigation of homicide in practice. The new claims of exclusive professional authority made by Lin Ji and other forensic reformers faced challenges when they encountered those of the other actors who were already involved in homicide investigation. Examining how the expertise of physicians was actually integrated into judicial procedure is thus essential for understanding the possibilities and challenges that reformers faced as they attempted to remake China's forensic practices within a legal context that tended to support a pluralistic arrangement of forensic knowledge and expertise in which judicial officials maintained significant authority.

5   Professional Politics of a Crime Scene

On the evening of January 15, 1924, a murder was committed in the densely populated area of the Outer City lying to the south of Qian Gate. Long a bustling commercial district, this part of Beijing was being eclipsed by Xidan and Wangfujing, new commercial centers that were rising to prominence under the economic and infrastructural changes that transformed the city during these years.[1] The killing took place in the home of Zhang Jiayi and her husband Shen Ruihong. Shen was a senior accountant at the Industrial Bank of Zhejiang, and Zhang, often referred to in the customary way as "Mrs. Shen née Zhang," was a normal-school graduate who had been a teacher until the previous year when the Wushan School that had employed her closed its doors. Working at the household was female domestic servant Liu, often simply referred to as "Liu Ma," who had been employed previously at Mrs. Shen's natal household. The police who arrived at the Shens' home on this particular evening found Liu Ma's body on the floor of the bedroom, bloody from a significant gash across the throat. While Mrs. Shen had been at home all evening, she could provide little information about what had happened to Liu Ma. Shen claimed that Liu had been attacked in the corridors of the residence by a person whom Shen herself never saw. Liu's violent encounter with this unknown person only became apparent when, wounded and bloody, she returned to the bedroom and sat down on the floor in front of Mrs. Shen. City procurators questioned this account, arguing instead that Liu had been killed after discovering Mrs. Shen's sexual affair with an unknown man, who was almost certainly also the murderer. The criminal case that they initiated would last for almost three years and move across all levels of the Republican court system.

At the center of the case was a crime scene covered in blood stains. With neither a confession nor witnesses other than Mrs. Shen, the procuracy's case hinged on this physical evidence, which included trails of blood droplets as well as bloody streaks, footprints, and fingerprints.

[1] Dong, *Republican Beijing*, 144–52.

136

According to procurators' interpretation of the evidence, Liu had been attacked in the bedroom, not the corridors of the compound as Mrs. Shen claimed. In their telling, the blood stains in the corridors had been fabricated by Mrs. Shen after the fact, as indicated by irregularities in the color and pattern of the stains. Most damningly, fingerprint evidence in the bedroom indicated that an unknown "third party" had been in the room around the time of the murder. Given that Shen had made no mention of another person in her account of what had happened, this suggested to procurators that she was lying. The procuracy's case was challenged by Shen's defense counsel, which put forward its own interpretation of the crime scene evidence; in the end, the prosecution was defeated upon appeal at the Capital High Court and Supreme Court. At each stage of the case, from the initial police investigation to the later appellate trials, the prosecution, defense, and judges mobilized a cadre of outside experts, including fingerprint examiners and medical school faculty, in their deliberations over the meaning of the forensic evidence.

Much was at stake in the involvement of physicians and other experts in the law, as we saw in the last chapter. An important question, though, is the extent to which experts such as Xu Songming, Sun Kuifang, and Lin Ji actually gained authority in the course of a criminal investigation and trial – traditionally the domain of police, procurators, and inspection clerks. That is, beyond the widely prevalent rhetoric of scientific authority and professional expertise, to what extent did professionals in medicine and other fields influence legal officials' understandings of the forensic evidence in actual cases as well as the standards that they used to assess the strengths and weaknesses of evidence more generally? Through a close analysis of the investigation of Liu Ma's murder and the legal proceedings that resulted, this chapter examines the different points at which outside experts became involved in this case, how they asserted their claims of professional knowledge and expertise, and how these claims were reconciled with those of the other groups that were already involved in the investigation of homicide.

Given that scientific evidence commonly enters modern legal proceedings amid a complex interplay of professional interests and rules of evidence, examining the role and motivations of these actors is important for understanding the process through which scientific knowledge gained authority in modern Chinese law.[2] In this case, both prosecution and

---

[2] For more on the ways in which the unique evidentiary procedures and professional interests that are at play in legal settings shape the form and content of scientific evidence, see Edmond, "The Law-Set." While Edmond is primarily concerned with legal proceedings in which expert witnesses provide evidence under oral questioning and cross-examination by both parties in front of a jury, his point that legal actors and procedures

defense had their own interests in interpreting the physical evidence in particular ways; likewise, each sought the assistance of outside experts in response to particular twists in the investigation and trial in order to support or attack particular interpretations and arguments. Moreover, the significant claims that procurators and other judicial officials made over body evidence in early twentieth-century China presented special challenges to physicians as they argued for the indispensability of their expertise in homicide investigations. In a judicial system that invested the initial inspection of a body with tremendous procedural importance, Chinese jurists were already equipped to make authoritative claims about forensic evidence early in an investigation and without the assistance of physicians. While this element of Chinese judicial procedure was criticized by proponents of forensic reform for hindering the progress of medical expertise in the law, it also had the effect of consolidating the authority of judicial officials over the dead body – one of the most important pieces of evidence in a homicide case. In this sense, examining how Chinese courts made use of outside experts reveals not simply the expanding influence of medical experts in the law, but a more amorphous inter-professional field of competing claims over physical evidence that emerged with the simultaneous rise of the modern legal and medical professions in early twentieth-century urban China.

The Early Investigation

Personnel of the Beijing police were the first to respond on the evening of Liu's murder. An officer patrolling Guanyin Temple Street came across Mrs. Shen, who had run shouting into the street, and quickly summoned other police officials as well as personnel from the Outer City Government Hospital, who rendered futile assistance to Liu Ma. The Shens' compound was a courtyard residence arranged along north–south and east–west axes. Walking into the main gate, which opened north on Guanyin Temple Street, one followed a walled passage leading to the compound entrance. Passing through this entrance brought one into the network of corridors connecting the compound's outward rooms with the central courtyard. In the southwest corner of the compound was the bedroom in which Liu's body was found. There were now blood stains in the corridor of the compound's inner entrance, including small footprints and several bloody streaks. A path of sprinkled blood led

establish the conditions in which scientific knowledge becomes actionable evidence (or, alternately, loses authority) pertains to the use of experts in Chinese courts during this period as well.

south from this corridor, ending in the bedroom. In this room too was a large amount of blood evidence: bloody fingerprints around the doorway, a washbasin filled with bloody water, and blood stains and droplets on the mattress and around the area where Liu's body was located.[3]

The Shens were soon taken to the district police office for questioning. According to Mrs. Shen's account, Liu had been keeping her company in the bedroom throughout much of the evening.[4] At one point, soon after Liu had left the room to get a kettle of water, Shen heard strange noises in the residence and what sounded like Liu entering into a confrontation with another person. After Shen called her back several times, Liu finally returned to the bedroom. Shen noticed at that point that Liu had blood on her nose and mouth and, thinking that she had simply become dizzy from a bloody nose, went over to wipe her face with water from the washbasin in the bedroom. It was at this point, Shen claimed, that she realized that Liu had a significant wound on her neck and ran out into the street with a lamp to summon the police. As the police report noted, Shen explained that the blood stains on her clothing had come from when she had tried to wipe the blood from Liu Ma's face.

On the following afternoon, procurator Wu Zehan arrived at the district police station, along with a secretary, forensic examiners Yu Yuan and Yu Delin, the "midwife" Ms. Xue, and judicial policemen, all of whom proceeded with police officials to the Shen household.[5] At the scene were a number of other officials as well as Liu Ma's son. As the forensic examination proceeded, the inspection clerks described the wounds on Liu's body in the standard terminology of the official examination form that was used in such cases. The written record of this examination would be copied and recopied in the legal documents of the case, constituting one of the least contested pieces of evidence:

We find an open laceration with purplish flesh starting just above and to the left of the throat and ending at a point to the left of the back of the neck. The wound is slanting and measures three *cun* and two *fen* [approximately 5 inches] long and two *fen* [0.3 inches] wide. It is deep enough to have reached bone at the back of the neck and damaged it. Death was caused by being cut with a blade.[6]

---

[3] This account of the police response and the crime scene is derived from early police reports: Report of Outer Right Second District, January 16, 1924, BMA J181-19-38290, 1–5; Report of Outer Right Second District, January 16, 1924, BMA J181-19-38290, 38–46; Report of officers of Criminal Affairs Office, February 1924, BMA J181-19-38297, 24–5.

[4] Report of Outer Right Second District, January 16, 1924, BMA J181-19-38290, 1–2.

[5] Ibid., 48–54.

[6] The record of the examination was copied in the various rulings associated with the case; i.e., Ruling of Capital High Court, August 2, 1926, BMA J65-4-381, 500. According to a commonly accepted conversion during this period, 1 *chi* = 14.1 inches. Accordingly,

In the weeks following the examination of Liu's body, investigators from the Beijing police and procuracy were mobilized. These included police detectives as well as members of the Judicial Department, the office subordinate to police headquarters that worked with the city procuracy to investigate crime and assist in the administration of criminal trials. As was part of its official mandate, the procuracy also played a direct role in this detective work, interrogating suspects and others who became implicated in the case as it unfolded. Representatives from these offices were present at the initial examination of Liu's body and soon busied themselves with tracking down leads across the city, examining the crime scene, and interrogating suspects.

Kitchen servants recently dismissed from employment at the house-hold were soon investigated, as was the husband of a teacher at the school at which Mrs. Shen had worked. A former employee of the school was located and questioned about this possible lead and, more generally, about Mrs. Shen's conduct while at the school. This was also the subject of an investigator's interview with a district policeman stationed near Mrs. Shen's own family's home. While none of Shen's former classmates could be located to vouch for her behavior as a student, the police officer himself testified to the good behavior of the family. By late February, procurators were investigating employees working in the kitchen of the Industrial Bank of Zhejiang, where Shen Ruihong worked. These included a kitchen servant who had noticed the commotion at the Shen household and notified Shen at the bank, as well as Sun Shoushan, the person who brought the meals packaged there to the Shen household. These officials subsequently questioned other employees of the bank regarding Sun's whereabouts on the night of the murder, and searched his belongings for blood-stained articles.[7]

By this point, in fact, the procuracy had already come to focus on Mrs. Shen as the target of suspicion. By the end of February, Shen had been taken into custody (on the same day that procurators ordered the release of kitchen employee Sun Shoushan, they denied Shen Ruihong's request for her release on bail).[8] Some degree of suspicion had probably existed since the beginning of the investigation. Even on the night that Liu

1 *cun* (1/10 *chi*) = 1.41 inches and 1 *fen* (1/10 *cun*) = 0.141 inches. H.G.W. Woodhead, *The China Year Book 1933* (Shanghai: The North-China Daily News & Herald, 1933), 596.

[7] Report of officers of Criminal Affairs Office, February 1924, BMA J181-19-38297, 24–32; Report of judicial police stationed at local procuracy, February 22, 1924, BMA J181-19-38297, 37–40; Report of judicial police stationed at local procuracy, February 25, 1924, BMA J181-19-38297, 45–8.

[8] Report of judicial police stationed at local procuracy, February 27, 1924, BMA J181-19-38297, 41–2.

was murdered, police authorities had noted apparent inconsistencies in Mrs. Shen's story. For example, they questioned her claim that she believed that Liu was simply dizzy from a bleeding nose after having heard what sounded like an altercation elsewhere in the compound – clear evidence that something was amiss.[9] Yet, as other leads dried up and all of the other persons who had been investigated were cleared, one suspects that a theory of the crime focused on Mrs. Shen herself would have become increasingly amenable to the investigating officials. These suspicions were buttressed by the crime scene, the most damning element of which was the fingerprint evidence in the bedroom.

Illicit Sex and Crime Scene Forensics: Building the Case against Mrs. Shen

The use of fingerprint identification was not uncommon in Beijing during this period, and one of the key figures in the early history of the use of the technique in China, an expert in fingerprint identification and police science named Xia Leibo (more commonly known by the name Quanyin), soon became involved in this case. Several years earlier Xia had been sent by the Beijing police department to study fingerprinting under the Shanghai Municipal Police, which would develop a reputation for maintaining one of the most lauded fingerprint agencies in the country. After his return to Beijing, Xia taught a specialization in fingerprint identification at the Advanced School for Police Officials (Jingguan gaodeng xuexiao) of the Ministry of Interior, a police academy established under the auspices of the central government in Beijing in 1917.[10] During the mid-1920s Xia also held the position of president of the Fingerprint Society (Zhiwen xuehui), a professional association comprised of graduates of the fingerprint program at the police academy as well as other interested individuals involved in law enforcement. Beyond pursuing these profession-building activities, Xia also provided fingerprint identification services to the Beijing police and private citizens, at times receiving glowing coverage in the pages of *Morning Post* and even in Shanghai's *Shenbao*.[11]

Early in the investigation into Liu's death, Xia Quanyin and Hui Hong, a police fingerprint examiner who had also received training in Shanghai,

---

[9] Report of Outer Right Second District, January 16, 1924, BMA J181-19-38290, 3–4. The procuracy raised this same point as part of its grounds for requesting preliminary investigation in the case. CB May 30, 1924, 6.

[10] SNB March 3, 1919, 11; SNB March 31, 1919, 11; SNB August 3, 1919, 11. For more on this institution, see Han and Su, *Zhongguo jindai jingcha shi*, 498–502.

[11] CB June 24, 1927, 6; CB June 29, 1927, 6; SNB October 3, 1925, 11.

photographed numerous finger impressions made in blood that were found around the bedroom and took prints from Mrs. Shen and Liu Ma's body.[12] Their initial focus was on five bloody finger impressions, four of which were photographed around the door of the bedroom and one of which was discovered near Liu's corpse. Xia Quanyin's initial analysis indicated that two of these did not match the fingerprints of anyone connected with the case – a finding that suggested that another person, whom Mrs. Shen had failed to mention in her account of what had happened, had been present in the bedroom at the time of Liu's murder.[13] Upon further examination Xia revised this finding, claiming instead that while one of these bloody finger impressions could be matched to Liu Ma, the other did indeed belong to an unidentified "third party" who had been in the bedroom and who had come into contact with Liu's blood.[14] This fingerprint – which was in actuality a partial impression of the surface of a finger containing a fraction of the ridge patterning contained in a full frontal or rolled fingerprint – would become one of the most important pieces of evidence in the case. The fingerprint evidence was damning because it suggested that Mrs. Shen had been lying about what happened that night. If the print of an unidentified person was discovered in the bedroom, the procuracy reasoned, then Shen's account of the killing had been contradicted. As explained by the Capital Local Court, which would find Shen guilty of aiding in the murder,

If not a strong and forceful man, one would not be able to cause [a wound as deep as Liu's]. Moreover, of the bloody fingerprints on and to the side of the door of the bedroom, there is one that Xia Quanyin has appraised as dissimilar to those of the defendant Mrs. Shen and Liu Ma. This is enough to prove that there was another person in the room who took part in the murder. The defendant [Mrs. Shen] was in the room at the time and cannot plead ignorance.[15]

The procurators' argument that Liu Ma had been murdered after discovering Mrs. Shen's affair with an unknown man also unquestionably relied on forms of suspicion that were the product of particular notions of normative gender roles and sexual performance, especially surrounding illicit sex. If Mrs. Shen's husband had discovered her in the act of an extra-marital affair and then killed her and her lover, for example, it is

---

[12] CB May 30, 1924, 6.

[13] Fingerprint examination report of Xia Quanyin, BMA J181-19-38290, 33–5.

[14] Xia's revised appraisal was published in the Fingerprint Society's publication *Zhiwen zazhi* (2 [1924]: *shiyan*, 1). The following issue (3 [1926]: *shiyan*, 41–2) contained another account of the case, also penned by Xia.

[15] Judgment of Second Criminal Court of Capital Local Court, May 29, 1925, BMA J65-4-382, 174.

plausible that judges would have found grounds for leniency – a reflection of long-standing assumptions about the immorality of women's extra-marital sex and the legal protection of a husband's violent response.[16] This was the world of social meanings and expectations in which the unidentified print became that of "a strong and forceful man" without any concrete proof of the attacker's identity. Evidence presented during the trial at the Capital Local Court lent credence to this theory. Mrs. Liu's daughter, who had also worked in the household, had observed that Mrs. Shen had received visits from male guests on several occasions – a highly suggestive piece of information that the judges of the Local Court accepted as supporting the prosecution's argument that Mrs. Shen had aided in the murder.[17] Shen's defense would later attempt to deflate the notion that a woman should be penalized for associating with males in this way, noting that this should not be considered strange during a time when "social intercourse has opened up."[18]

That the procuracy's case was flimsy is suggested by the fact that in late May 1924, the Capital Local Court in Beijing refused to allow prosecution of Mrs. Shen to proceed. This decision was made during the preliminary investigation phase, a legal procedure referred to as *yushen* in Chinese. This process – which had been derived from Japanese and, before it, French law – involved the review of case evidence by designated procurators and judges for the purpose of deciding whether or not to proceed with prosecution and trial.[19] Part detective work and part preparation for later phases of a trial, this procedure sat uneasily at the intersection of the authority of procurators, who were already responsible for investigating crimes and initiating prosecution, and judges, who were themselves invested in determining the facts of a case in conjunction with the procuracy. In this case it was Judge Tian Chou (whom we previously encountered pronouncing the significance of wounds on "vital spots" in Chapter 2) who decided that the state of the evidence was insufficient and refused to allow the prosecution of Shen to go forward.[20] In his

---

[16] As Jennifer Neighbors shows, Republican judges maintained the prerogative to apply leniency in cases involving husbands committing murder following the discovery of adultery. Jennifer M. Neighbors, "The Long Arm of Qing Law? Qing Dynasty Homicide Rulings in Republican Courts," *Modern China* 35, no. 1 (2009): 23–7.

[17] Judgment of Second Criminal Court of Capital Local Court, May 29, 1925, BMA J65-4-382, 176.

[18] Requests of Ye Zaijun to Capital High Court, January 20, 1926, BMA J65-4-381, 323.

[19] For more on the *yushen* procedure, see Huang Yuansheng, "Jindai xingshi susong de shengcheng yu zhankai: Daliyuan guanyu xingshi susong chengxu panjue jianshi (1912–1914)" [The emergence and development of modern criminal litigation: Judgments and commentary of the Supreme Court regarding criminal litigation procedures (1912–1914)], *Qinghua faxue* 8, special issue: Research on codification (2006): 101–6.

[20] CB May 30, 1924, 6.

response to the procuracy's protest of this decision, Tian wrote that the idea that Liu had been murdered in the bedroom by Mrs. Shen and an unidentified killer was "purely wishful thinking without any actual proof" – a strong indictment of the procuracy's case and the evidentiary claims on which it was based.[21] Yet, in the end, the procuracy successfully appealed Tian's decision at the Capital High Court, which ordered another preliminary investigation to definitively resolve the question of Shen's involvement in the killing.[22]

Despite the considerable doubts that this case raised among some Beijing jurists, in the trial that followed the Capital Local Court found Mrs. Shen guilty as an accomplice in the murder.[23] Not only did the Capital Local Court largely ignore the High Court's questions about the evidence, but it straightforwardly accepted the procuracy's narrative of Liu's murder. In the objective-sounding "facts" (*shishi*) and "grounds" (*liyou*) on which the judgment was made, the court simply recapitulated the procuracy's case: Liu Ma had discovered Mrs. Shen's illicit sexual affair; Shen and the unnamed male agreed to kill Liu; the murder took place in the bedroom; following the killing, Mrs. Shen used blood from Liu's wounds to create false footprints and bloody marks in the corridor in order to support her story.[24] These claims were based on the judges' interpretation of several key pieces of physical evidence, of which two were especially important. One was a thick blood stain with two trails of blood spray found in front of the bed, which court officials had personally inspected during a visit to the crime scene.[25] This stain, seemingly from the blood that jetted out of Liu Ma's throat, indicated that, in their estimation, she had been attacked in the bedroom and not in the corridor. The other crucial piece of evidence was the partial fingerprint examined by Xia Quanyin.

Re-examining the Evidence

If not for Mrs. Shen's defense counsel, the decision of the Local Court might have stuck. The institutionalization of a formalized and legitimate

---

[21] CB June 16, 1924, 6; Opinion of Judge Tian Chou, June 5, 1924, BMA J65-4-383, 16.

[22] Ruling of Capital High Court, June 21, 1924, BMA J65-4-383, 46–55.

[23] Judgment of Second Criminal Court of Capital Local Court, May 29, 1925, BMA J65-4-382, 168–78.

[24] The explicit presentation of "facts" and "grounds" in legal judgments was, as Michael Ng notes, a new practice that was the result of early twentieth-century judicial reform. As Ng also notes, this new format could accommodate older forms of judicial knowledge, such as a record of the forensic examination. Ng, *Legal Transplantation in Early Twentieth-Century China*, 14–7.

[25] Record of examination of crime scene, May 19, 1925, BMA J65-4-382, 120.

role for defense lawyers at trial was one of the innovations of the legal reforms carried out in China during the first decades of the twentieth century. As Eugenia Lean has vividly demonstrated through the case of Shi Jianqiao, who received a pardon after assassinating the former militarist Sun Chuanfang in 1935 in order to seek revenge for her father's execution, the crafting and implementation of an effective defense strategy could make all the difference in highly public trials.[26] In that case, the defense skillfully appealed to public sympathy on the basis of Shi's morally and ritually sanctioned motives for the killing. In this case as well, we find an expert defense able to mobilize varied resources on behalf of their client. Mrs. Shen was represented by Liu Chongyou (1877–1941) and Ye Zaijun, highly skilled lawyers who proved more than capable of engaging the procuracy's case point by point. A graduate of the Capital School of Law and Politics (Jingshi fazheng xuetang) established in the last years of the Qing, Ye Zaijun would go on to hold a number of positions over the course of his career in the Capital Local and High courts as well as in the Supreme Court under the Nationalists.[27] Liu Chongyou was a graduate of Waseda University who participated in a number of the representative political bodies established under late Qing constitutionalism and in the early years of the Republic. By the mid-1920s, Liu had achieved a very public profile for having defended a number of students, including Zhou Enlai (1898–1976), who had been arrested for political activism in the wake of May Fourth.[28]

Following their appeal to the High Court after the trial in the Capital Local Court, the defense made a series of requests to the appellate judges to re-examine the forensic evidence in the case.[29] One request was for experts in legal medicine to examine the scene of the death and the blood evidence. This was the core of the case against Mrs. Shen; if it could be proven that Liu had been fatally wounded in the corridor and not the bedroom, then the case against her would evaporate. In calling for med-ico-legal experts to re-examine this evidence, the defense was making the claim that only specialists trained in the relevant fields of science, not officers of the court, were equipped to make claims about its meaning – the same argument that forensic reformers such as Lin Ji were making

---

[26] Lean, *Public Passions*, 115–24.
[27] Woodhead, *The China Year Book 1933*, "Who's Who," 468; Liu Guoming, *Zhongguo Guomindang bainian renwu quanshu* [Biographical Compendium of Personages from One Hundred Years of the Nationalist Party] (Beijing: Tuanjie chubanshe, 2005), 366.
[28] For an overview of the wide-ranging political and legal activities with which Liu Chongyou was involved during this period, see Liu Guangding, *Aiguo zhengyi yi lüshi: Liu Chongyou xiansheng* [A patriotic and justice-seeking lawyer: Mr. Liu Chongyou] (Taipei: Xiuwei zixun keji gufen youxian gongsi, 2012).
[29] Requests of Ye Zaijun to Capital High Court, July 13, 1925, BMA J65-4-381, 182–90.

during this same period. Of course, the defense had a clear interest in shifting the authority to interpret the evidence away from the procuracy, which had put forward an interpretation of the crime scene that was damaging to their client. Moreover, while Shen's defense demanded the involvement of outside experts, at times they engaged directly with the scientific questions at the heart of the case – the very issues that were supposed to fall under the exclusive purview of the experts.

Less than a week after the defense counsel's request, the Capital High Court sought expert appraisals from faculty of the National Medical College in Beijing. Faculty from this school as well as from PUMC had already been involved in the case during the preliminary investigation. Their examination of various fluids found at the scene of Liu's death largely confirmed Shen's account of what had happened and provided part of the grounds for Judge Tian Chou's initial ruling against going forward with prosecution.[30] At this later stage of the case, the National Medical College was contacted again for a re-examination of the blood evidence. This new appraisal was handled by Xu Songming, who was professor of pathology and forensic medicine at the time, as well as Lin Zhen'gang and Xu Shijin, graduates of the medical school who had been kept on as assistant instructors in pathology. On the morning of July 25, 1925, a party comprising High Court judges, procuratorial authorities, Ye Zaijun, Mrs. Shen and her husband, and the medical school faculty went to the Shens' residence for another walk-through of the crime scene.[31] The medical school faculty retested samples taken from walls and various objects inside and outside of the bedroom.[32] The examination protocol that they followed to confirm the presence of blood utilized some of the most advanced forensic techniques available at the time, including the Japanese legal medicine expert Takayama Masao's test for confirming blood stains on the basis of hemochromogen crystals produced after treatment with a pyridine-containing reagent.[33]

This appraisal of the blood evidence did cast doubt on the procuracy's argument that Liu had been killed in the bedroom. Xu and his colleagues directly addressed this question in one of the expert opinions that they

[30] CB May 30, 1924, 6.
[31] Record of examination of crime scene, July 25, 1925, BMA J65-4-381, 207–20.
[32] Expert opinion of Xu Songming, Lin Zhengang, and Xu Shijin, July 29, 1925, BMA J65-4-381, 238–41; Expert opinion of Xu Songming and Xu Shijin, December 24, 1925, BMA J65-4-381, 289–92.
[33] Published in 1912, Takayama's procedure would be widely used by forensic examiners across the world throughout the rest of the twentieth century. For more on the history of Takayama's test, see R.E. Gaensslen, *Sourcebook in Forensic Serology, Immunology, and Biochemistry* (Washington: U.S. Department of Justice, National Institute of Justice, 1983), 86–7.

submitted to the High Court. Discussing the large blood stain in the bedroom that, it had been presumed, indicated where Liu had been attacked, they noted:

Moreover, there is a blood stain with spatter slanting to the northeast at a spot on the floor several *chi* [1 *chi* = approximately 14 inches] from the bed. It seems to be the stain from blood spurted out at the time that Liu was wounded, but one cannot determine that it was in the bedroom that she was wounded on the basis of this alone. While we inspected the blood stains on the other spots, it is difficult to infer the location at which Liu was wounded (whether it was the bedroom or the corridor).[34]

Ultimately, Xu's finding that it was impossible to tell where Liu had been attacked on the basis of the blood stain raised fundamental questions about the prosecution's case. The fingerprint evidence too had been essential for the case against Mrs. Shen by suggesting that another person had been in the bedroom on the night of the murder. During this phase of the trial, this evidence was re-evaluated by an expert in fingerprinting named Qian Xilin, who had studied in Berlin and held an honorary position in the Fingerprint Society with which Xia Quanyin himself was affiliated.[35] In late December 1924, Qian submitted a report that questioned Xia's claim that the bloody partial print belonged to an unidentified "third party," arguing instead that its impartial nature and lack of clear ridge characteristics precluded any valid identification at all.[36] In a separate report, Qian further criticized the process through which the original comparison had been made between the unidentified print and the exemplar prints obtained from Mrs. Shen, Liu's corpse, and the others involved in the case.[37] These samples were frontal impressions of the fingers, not rolled prints that included the sides of the fingers in addition to the front. Given that the bloody print had been left by the side of a fingertip, Qian argued, it was impossible to establish a valid comparison because this region of the finger had never been recorded on the exemplars. This would become a fatal challenge to the fingerprint evidence in the case: because a new set of rolled fingerprints could not be obtained from Liu's long-interred corpse, it would be impossible to claim with certainty that the partial print was not in fact hers.[38]

[34] Expert opinion of Xu Songming, Lin Zhengang, and Xu Shijin, July 29, 1925, BMA J65-4-381, 238–9.

[35] *Zhiwen zazhi* 2 (1924): *fulu*, 15.

[36] Expert arguments of Qian Xilin, December 24, 1924, BMA J65-4-381, 31–4.

[37] Expert opinion of Qian Xilin, February 2, 1925, BMA J65-4-381, 337–9.

[38] Qian's arguments prompted Xia Quanyin to submit another expert opinion in which he stated definitively that he had in fact checked the print in question against the rolled prints of everyone involved in the case, yet was unable to do so for Liu's fingers. Xia had only taken frontal impressions of these because, he claimed, the fingers had been tightly

In the end, it was on the basis of these new expert opinions that the case against Mrs. Shen was overturned. Xu's statement about the indeterminacy of the blood evidence would be cited in the High Court's judgment, much as it would be in the Supreme Court's own exoneration of Mrs. Shen after the High Court's trial.[39] The courts also cited Qian's discussion of the problems of the partial print, noting that "what Qian Xilin says is not without grounds. As such, this fingerprint is insufficient as proof that another person was in the bedroom."[40] In accepting the experts' assessments of the blood and fingerprint evidence, judges of the appellate courts had effectively delegated the authority to interpret these crucial areas of evidence to them. They placed trust in the experts' professional credentials and accepted their immediate claims over the evidence as well as the authority of the knowledge on which these were based. In the process, an injustice had seemingly been averted.

Science for Lawyers

In fact, the implications of this case for the authority of experts in the law are less clear than they might seem. Looking closely at the ways in which the prosecution, defense, and judges utilized the evidence of the experts presents a picture not of the absolute and exclusive authority of the latter, but, rather, of practical negotiations in which legal professionals themselves made decisions about when and how to use such evidence in support of their own arguments and judgments. Indeed, it is important to remember that when Xu Songming and Qian Xilin presented evidence as outside experts, they did so in an institutional setting in which the law's own professionals already managed many if not all aspects of what happened in the course of a trial. Judges, procurators, and defense lawyers maintained their own authority over legal proceedings on the basis of their professional qualifications and experience as well as the procedurally defined roles that they played in judicial process. While outside experts could make authoritative claims over the circumscribed areas of evidence in which they were recognized to be specialists, legal officials held a substantially broader mandate to deliberate over multiple forms of

---

clutched and could not be easily manipulated. Expert opinion of Xia Quanyin, May 25, 1926, BMA J65-4-381, 408–10.

[39] Judgment of Capital High Court, August 2, 1926, BMA J65-4-381, 505. The Supreme Court also cited Xu's evidence. CB January 6, 1927, 6.

[40] Judgment of Capital High Court, August 2, 1926, BMA J65-4-381, 509–11. In its own ruling, the Supreme Court too cited Qian's discussion, using it as part of its own argument about the insufficiency of the evidence in the case. CB January 6, 1927, 6.

evidence while synthesizing the most compelling facts as the basis for their interpretations and arguments.[41]

These negotiations over the authority to decide the facts of the case played out around the expert appraisals, known in Chinese as *jianding*, through which Xu Songming and others presented evidence to the court. This legal procedure for submitting an expert's opinion regarding particular questions of evidence had been derived from Japanese law, where it was called *kantei*, and, before it, from the German practice of submitting expert opinions into law, known as *gutachten*.[42] This conception of the outside expert's role in the law implied a new division of epistemological labor: judicial officials were to accept that their knowledge was limited when it came to questions of science, technology, medicine, and other fields. By implication, obtaining reliable evidence required seeking out individuals with specialized knowledge and expertise. As Zhang Yuanjie, a Qing diplomatic official to Japan, noted in his preface to one of the early translations of Japanese legal medicine,

As the knowledge of a nation's people develops, affairs within society become increasingly complex and legal questions gradually become more complicated. Not all difficulties can be resolved by resorting to practices that are commonly known. Instead, those with special ability must appraise the matters. Judges undoubtedly know the law, yet when examining evidence it is impossible to attain fair legal judgments when one is without the assistance of those who have other kinds of special ability. For example, the usual practice of all countries is to make testing for poisons the specific duty of a chemistry technician or pharmacologist. When encountering a case of suspected poisoning, an expert in legal medicine only appraises whether or not it was a case of poisoning and has a chemist or pharmacologist examine the nature of the toxin.[43]

In the modern legal practices that Zhang was describing – implicitly those that Japan and countries of continental Europe had already adopted – a high level of specialization had been achieved in the law. Not only did judicial officials rely on experts in legal medicine, trained physicians who had already mastered a large body of specialized medical knowledge of relevance to legal questions; they also sought the assistance of those with

---

[41] On this point, see Michael Ng's discussion of the new conceptions of legal proof introduced during the New Policies judicial reform, which explicitly encouraged judges to consider multiple forms of evidence (instead of emphasizing oral confession) while giving them the authority to decide how to weigh the different sources of evidence in a particular case. Ng, *Legal Transplantation in Early Twentieth-Century China*, 13, 20–2.

[42] For an overview of the procedures for expert testimony in Germany, see American Association for the Advancement of Science, *Reports on the Use of Expert Testimony in Court Proceedings in Foreign Countries* (Washington: Press of Byron S. Adams, 1918), 142–50.

[43] Wang You and Yang Hongtong, *Shiyong fayixue daquan* [Great compendium of practical legal medicine] (Tokyo: Kanda insatsujo, 1909), *xu* 2–3.

deeper expertise in toxicology or other fields, when the circumstances of a case demanded it. This notion of the limits of legal professionals' knowledge vis-à-vis those of specialists reflected broader societal shifts in patterns of occupational expertise that were connected to the rise of the professions and the increasing complexity of the fields of technical knowledge on which more and more areas of the economy, governance, and social relations were based. Forensic investigation was only one area of judicial practice that touched on technical matters that lay beyond the expertise of judges. There were many other situations as well in which jurists might need the assistance of experts when adjudicating matters pertaining to the new forms of scientific knowledge and industrial technology that were rapidly transforming society and economy.[44]

Such expert opinions were different from the use of expert witnesses in Anglo-American courtrooms. Since the late eighteenth century, experts in Britain and the United States had participated in legal proceedings primarily as witnesses, thus giving them the perceived status of partisans called by either party and making them subject to contentious cross-examination before the court.[45] By contrast, in the legal procedures of late nineteenth-century Germany, from which Japanese and Chinese legal reformers had derived significant elements of their countries' modern legal codes, experts submitted opinions under the direction and authority of the court itself, and the presiding judges played an active role in requesting and utilizing these expert opinions. As in this case, deliberations over evidence could take place through judges' reading of carefully composed written documents rather than the forms of cross-examination that, as expert witnesses in the United States and Britain were aware, could deeply challenge an expert's credibility.[46]

The written appraisal reports (jianding shu) that experts submitted to Chinese judicial authorities were thus of the utmost importance. Such reports could take many forms, ranging from a brief, informal description of an examiner's findings to a detailed report written in a standard format explaining the relevant scientific principles in addition to the evidence at

---

[44] This point was made, for example, in another translation of Japanese legal medicine co-authored by the prolific translator Ding Fubao (1874–1952). Ding Fubao and Xu Yunxuan, Jinshi fayixue [Modern legal medicine] (Shanghai: Wenming shuju, 1911), 2–3.

[45] Tal Golan, Laws of Men and Laws of Nature: The History of Scientific Expert Testimony in England and America (Cambridge: Harvard University Press, 2004).

[46] For the challenges that physicians faced when serving as expert witnesses in nineteenth-century American courts, for example, see Mohr, Doctors and the Law, 94–108, 197–212. Mohr identifies the overwhelmingly negative experience of physicians in court as one of the factors that contributed to the declining fortunes of medico-legal institutional development in late nineteenth- and early twentieth-century America.

hand. The latter characterized the reports of Lin Ji, whom we encountered in the last chapter as one of the early proponents of forensic reform in China. Every one of the written appraisals that Lin Ji sent to judicial authorities during the 1930s included an institutional cover page, a restatement of the query of the authorities who had sought the expert opinion, a detailed description of the physical objects prior to testing, and an explanation of the results of the testing as well as any relevant scientific principles, experiments, or other points of interest that had informed the analysis. From the perspective of the experts themselves, the goal of writing these documents was to establish their professional authority while making highly technical findings intelligible to legal officials who were not specialists. In this sense, written appraisals can be understood as a form of popularized scientific knowledge – one targeted not to a mass audience, but to legal professionals, and meant to be used in the specific context of legal decision-making.[47]

The significance of an expert appraisal could be quite different when viewed from the perspective of judges and procurators, for whom it was simply one source of evidence over which they could deliberate. Judicial officials themselves had the exclusive authority to decide whether or not to utilize a particular expert opinion and when to seek out additional examinations of the evidence. As we will see in Chapter 6, for example, the physician's autopsy was never established as a legally mandated procedure of policing or forensic investigation; whether or not an autopsy was carried out was a question left to the discretion of judicial authorities, an area of latitude over which forensic reformers struggled for influence. When expert appraisals were used, the knowledge contained within became accountable to rules and procedures of admissibility as defined in the regulations on criminal and civil procedure. These laws defined who could submit an expert opinion, how they were to do so, and what form it had to take.[48] Not following the proper procedures could render an expert opinion invalid, regardless of the credentials of the examiner or the scientific status of the knowledge itself. In this case, for example, the results of Xu Songming's testing did not have a legal effect until they were accepted as evidence by the judges tasked with trying the case. This occurred when the High Court judges used Xu's expert opinion in their final judgment, having been convinced that the evidence underlying the prosecution's case was not persuasive. In the end, the judges were the

---

[47] Cf. Schmalzer, *The People's Peking Man.*

[48] At the time of Mrs. Shen's trial, one set of relevant regulations was contained in Volume 1, Section 8 of the "Criminal Procedural Regulations" (Xingshi susong tiaoli), promulgated in 1922. *Sifa gongbao*, issue 164, supplementary issue 27 (1922): 30–2.

ones to decide which evidentiary facts would inform the final verdict, not the outside experts.

In these ways, the scientific knowledge contained in an expert's appraisal only gained authority in the law when it was accepted by those who were already invested with authority in this unique institutional and procedural context. And, once submitted, the larger implications of an expert's appraisal were also subject to negotiation by the legal professionals who deliberated on and ultimately decided which facts were the most important. Understandably, this process was informed by the various interests and patterns of argumentation that shaped the course of a trial more generally. This is apparent, for example, in the defense's use of Xu Songming's appraisal to criticize the prosecution's interpretation of the crime scene evidence. In the arguments that they made before the Supreme Court, Shen's defense discussed the blood evidence of the case in detail, identifying this as the key to the prosecution's argument that Liu had been attacked in the bedroom and not the corridors of the house.[49] In addressing this evidence, the defense drew on *Scientific Methods for Investigating Homicide Cases* (Satsujin kagaku-teki sōsahō), a primer on criminal investigation penned by the well-known Japanese procuratorial official Nanba Mokusaburō, who wrote popular books on criminal procedure and forensics that went through numerous editions in Japan and were even translated into Chinese.[50] Prior to even discussing Xu Songming's appraisal, the defense extracted from this book several general principles relevant to the examination of blood stains. These were the principles that (A) blood stains can exhibit many different colors depending on their age and other environmental factors; (B) the movements undertaken by a wounded person can create all different kinds of blood patterns on the floor and other objects when droplets are emitted from the body and land on these surfaces; and (C) that blood stains can appear not simply at the site at which a victim was wounded, but at other locations to which the victim subsequently moved.

The defense then used these principles of blood analysis, which they themselves had extracted from Nanba's book, to systematically criticize the forensic evidence that had been used to support the prosecution's case. First, they addressed the claims made by the prosecution and Capital Local Court in the first trial that the presence of lighter and darker

---

[49] The defense's arguments were published in their entirety in *Morning Post*: CB December 7, 1926, 6; CB December 8, 1926, 6; CB December 9, 1926, 6; CB December 10, 1926, 6; CB December 13, 1926, 6; CB December 15, 1926, 6; CB December 17, 1926, 6.

[50] Nanba Mokusaburō, *Fanzui soucha fa* [Methods of criminal investigation], translated by Xu Suzhong (Shanghai: Shanghai faxue bianyi she, 1933).

blood stains and pointed droplets in the corridors of the compound indicated that Mrs. Shen had sprinkled bloody water from the wash basin in order to fabricate evidence that Liu had been attacked outside of the bedroom. These claims were untenable, the defense argued, given the principles (A) and (B), which could explain away a wide range of colors and droplet-shapes as perfectly normal variations in the appearance of blood stains.[51] By implication, only a trained examiner, well-versed in legal medicine, could accurately interpret blood stains on the basis of the varied physical appearances that they could take on in relation to different surfaces, movements, or other environmental factors. Without a more specialized understanding of these different variables, they suggested, this would be a futile task. The defense then used principle (C), that the blood emitted from a wound can land at any location to which a wounded person moves, to address the Local Court's claim that the significant blood stain in the bedroom was evidence that Liu had been attacked there and not in the corridor.[52] It was plausible, they suggested, that she had been attacked in the corridor and subsequently walked to the bedroom. It was at this point that the defense invoked Xu Songming's expert opinion to support the arguments that they were already making about the blood evidence. Quoting the key passage in Xu's report, the defense noted:

The expert opinion submitted on July 29th, 1925 by Professor Xu Songming of the Beijing medical school stated: "There is a blood stain with spatter slanting to the northeast at a spot on the floor several *chi* from the bed. It seems to be the stain from blood spurted out at the time that Liu was wounded, but one cannot determine that it was in the bedroom that she was wounded on the basis of this alone." This is an expert opinion that was made in accordance with principle C, and those mistaken assumptions that were made about the blood evidence from the start should be all the more clear.[53]

In fact, the defense had been advancing an analysis of the blood evidence that went beyond the relatively conservative opinions of Xu, who had made no mention in his report of the "principle C," which the defense themselves had obtained from reading Nanba's *Scientific Methods for Investigating Homicide Cases*. Moreover, Xu's opinion was not that the blood evidence in the bedroom proved that Liu had not been attacked there, but that one could not make *any* claims about where Liu had been attacked on the basis of the blood stain. By claiming that Xu was, in fact, talking about "principle C," the defense was extending his findings in a direction that was advantageous to their client. They were also subtly

[51] CB December 8, 1926, 6.   [52] CB December 9, 1926, 6.   [53] Ibid.

challenging Xu's presumed monopoly over the authority to interpret the blood evidence.

In this case, the defense counsel emerged as one of the strongest proponents for medico-legal experts' involvement in the law. Their accusation that police and legal officials' arguments about the blood stains represented nothing more than "unscientific conjecture," for example, resonated with the arguments of those who were pushing for greater involvement of physicians in legal cases and a transformation of the basic epistemological norms underlying jurists' engagements with forensic evidence.[54] Yet, when we examine how the defense engaged the legal medicine experts' evidence in practice, we find that they themselves played a significant role in establishing the meaning of the evidence for the case at hand. In this way, the authority of an expert such as Xu Songming was contingent on legal professionals' utilization of his specialized knowledge within the procedures of investigation, debate, and deliberation that already undergirded legal decision-making. The fate of legal medicine and other fields of extra-legal knowledge in court proceedings was thus determined by the varied interests that were already at play within this unique institutional setting.

Missing Bodies, Redundant Professionals, and the Limits of Medical Expertise in the Law

Chinese procurators' use of their own subordinate body examiners also shaped when, how, and for what purposes outside experts became involved in homicide investigation. Early in this case, city procurators oversaw a forensic examination of Liu Ma's body conducted by their own forensic personnel, who inspected and documented the fatal wound. As we have seen, this pattern was widely followed in the investigation of homicide in Beijing and other parts of China. The fact that inspection clerks were usually the only forensic examiners used in the early stages of an investigation had deep implications for the prospects of physicians to become involved in the case. In this instance, Xu Songming and his colleagues were deployed at a later stage of the case, at a time when many questions of forensic evidence had already been decided by procurators and the police. Once Liu Ma's body had been examined, it appears as though procurators and judges saw no need for the medical school faculty to re-examine the bodily evidence or to have any engagement with the corpse at all. Experts in pathology such as Xu Songming could offer highly detailed examinations of internal organs, bone, tissues,

---

[54] CB December 8, 1926, 6.

and cells, in the process examining the body at scales that were many times smaller than those that inspection clerks could even see with the naked eye. Yet, in a case such as this one, procurators and judges were satisfied with the inspection clerks' confirmation of the seemingly self-evident fact that Liu Ma had died at the hands of a blade-wielding assailant.

In this case, the fact that the medical school experts were not involved from the beginning of the investigation also meant that their access to the physical evidence had been compromised. When the medical school faculty accompanied the entourage of judges and others to the crime scene in late July 1925, they found that when the covers protecting the blood stains in the corridor were removed, this evidence had degraded over time. Much the same was found of the blood evidence in the bedroom.[55] In the written appraisal that they submitted to the Capital High Court, Xu and his colleagues explained that their examination of the remaining blood evidence would make few substantive contributions to judicial authorities' understanding of what had happened in the case:

Given that we never examined the corpse in this case and that in this round of examination the blood stains in the inner entrance corridor and in the compound had already diminished because of the passage of time, and also because the objects in the bedroom had been moved around, it is especially difficult to make an appraisal. Therefore we cannot provide satisfactory answers for the questions that the court has asked.[56]

Thus, the delayed entry of physicians into the investigation process could have implications for their effectiveness and, ultimately, their professional authority. Xu Songming's experience in this case was not unusual. Legal medicine experts tended to become involved in cases after a dispute or other complication had arisen regarding the evidence. It was less common for them to handle everyday investigations of death of the kind that constituted much of the procuracy's forensic examination work in Beijing. By the time that outside experts were called upon to examine the evidence, much investigation had already taken place, the body and other evidence had been inspected, and judicial authorities had decided on the basic questions that would guide the case.

Ultimately, as this case demonstrates, building a compelling narrative of what happened in a crime required much more than just the physical evidence, not to mention that smaller body of physical evidence examined with forensic techniques associated with the special procedures of

[55] BMA J65-4-381, 209–12.
[56] Expert opinion of Xu Songming, Lin Zhengang, and Xu Shijin, July 29, 1925, BMA J65-4-381, 240.

laboratory science. The involvement of Xu Songming and other experts in the later phases of the case did not provide a new understanding of who killed Liu Ma or the motives that were involved. They simply cast doubt on earlier interpretations of the physical evidence. Investigating motives and alibis could be just as important as examining the physical evidence and, as we have seen, doing so relied on commonsensical assumptions about what counted as normative social behavior in the time and place of early twentieth-century Beijing. The evidentiary facts that formed the basis for legal decision-making tended to rely on forms of reasoning and standards of proof associated with police detectives and legal officials, arguably the Republican court system's most common sources of knowledge about criminal acts. The interrelated questions of how to provide useful context to scientific evidence and productively reconcile the knowledge-claims of experts, police, and legal officials are ones that shape the limits and possibilities of forensic expertise in many modern legal settings.[57] Judicial officials in Republican China addressed these challenges with the resources that they had, including the body examiners who already staffed their offices. The forensic techniques of the latter had long provided officials in China with ways of reconciling bodily signs, the social context of homicide, and the legal categories that defined notions of criminal responsibility.

The challenges that physicians faced in persuading judicial authorities to utilize their expertise in homicide investigation thus reflected the fact that the pursuit of scientific truth was not, and could not have been, the overriding goal of this process. Judicial officials were undoubtedly invested in discovering what had happened in a case such as this one. As this case has suggested, however, obtaining this knowledge relied as much on investigation leads and social assumptions as it did on laboratory tests. In fact, as much as proponents of forensic reform might argue for the special status of their knowledge vis-à-vis that of legal officials, one cannot say that the epistemological capabilities of the latter were generally viewed as deficient in China during this period. As Michael Ng has shown, jurists claimed to represent the "facts" (*shishi*) of a case in their judgments and to base their decisions on this standard of proof.[58] This common idea was itself a powerful claim of epistemological and professional authority, one that would have resonated with the broader political and social significance of this modern epistemological category. As Tong Lam has demonstrated in his study of social surveys and state-building, for example, this new

---

[57] Timmermans, *Postmortem*; Thomas Scheffer, "Knowing How to Sleepwalk: Placing Expert Evidence in the Midst of an English Jury Trial," *Science, Technology, & Human Values* 35, no. 5 (2010): 620–44; Lynch et al., *Truth Machine*, 190–219.
[58] Ng, *Legal Transplantation in Early Twentieth-Century China*, 17, 20–2.

conception of empirical knowledge played an important role in legitimizing the modern Chinese state and its interventions into a newly imagined social body.[59] Even if jurists did not base their judgments on scientific knowledge, they did routinely invoke other conceptions of empirical knowledge that carried their own epistemological weight.

Similarly, while inspection clerks might be criticized for using practices that did not accord with the epistemological standards of modern laboratory science, their forensic claims about the body were not commonly denied the status of being valid evidence – a point that is amply demonstrated by judges' common practice of simply copying the record of the forensic examination in the "facts" (*shishi*) section of their legal judgments, much as they did in this case. In this way, inspection clerks' claims about body evidence were routinely associated with this category of authoritative knowledge despite the fact that their knowledge could appear quite uncertain when evaluated according to the epistemological standards of laboratory science. This suggests that for Republican jurists the scope of valid evidentiary "facts" was broader than that which was exclusively associated with scientific knowledge, even when physical evidence pertaining to the body itself was concerned. Ironically, there is a sense in which inspection clerks' forensic claims about the body gained authority precisely because they were not constructed as exclusive knowledge that was only accessible to a specially trained group of experts. Their body examinations often carried an assumption of transparency: the supervising procurator (not to mention others present at the scene) was expected to be able to understand the forensic claims that were being made by the inspection clerk.[60] By implication, one might say that inspection clerks did not have to work as hard to convince others of the validity of their knowledge-claims, which were not construed as being fundamentally distinct from the "facts" that could be discovered by judicial officials in the course of an investigation or trial. It was precisely this assumption that the meaning of physical evidence was transparent to non-specialists that forensic reformers targeted in their attempts to convince judicial officials to give up their claims to interpreting forensic evidence without the assistance of outside experts.

Conclusion

Inherent in the use of outside experts in early twentieth-century China was a new vision of societal division of labor as well as a new set of professional politics in the law. For physicians, medico-legal specialists, and members of

---

[59] Lam, *A Passion for Facts*.
[60] Chang, "'Zhongguo chuantong fayixue' de zhishi xingge yu caozuo mailuo," 14–5.

other nascent professions, getting legal courts to accept their expertise was important for the process of establishing their authority, prestige, and economic livelihood in Republican society. That legal officials did delegate certain areas of fact-finding to the experts reflected the new assumption that professions beyond the law should have a measure of legal decision-making authority. At the same time, the point of seeking an expert appraisal was to facilitate judges' own deliberations over the facts, not to give the outside experts the formal legal authority to render judicial decisions. It was precisely the modern legal profession's role in overseeing court cases, settling disputes, rendering justice, and administering the machinery of the state that constituted the basis for its own professional status and authority. Thus, while the notion that a modern legal system needed the expertise of outside specialists was accepted early on, the boundaries of the experts' interpretive authority over evidence had to be worked out with that of the law's own professionals. It is in the interplay of these competing forms of knowledge and authority that this chapter has located the early history of modern expert evidence in Chinese courtrooms.

In this case and others, Chinese courts sought out the assistance of pathologists, medico-legal experts, and other evidence specialists. Yet, as this case has also demonstrated, the scope of experts' involvement was often shaped by legal officials' own professional interests in a case. That is, judicial authorities decided when to utilize outside expertise, which evidence to send for examination, and which questions to ask of the evidence. Because, in many instances, judicial officials had already over-seen their own examination of the body, the involvement of physicians could be quite peripheral within an investigation. Thus, in this case, while Xu Songming and the faculty of National Medical College played a crucial role in closing the case and averting a potential injustice, their role was also circumscribed. Moreover, the involvement of physicians or other experts did not necessarily challenge the authority of the *Washing Away of Wrongs* or the various examination practices associated with it.

If the picture that has emerged from this chapter is one of the ambivalent influence of medical professionals in homicide investigation, it is important to not equate this with ambivalence toward modern Western science more generally. The intellectual concepts and technical knowledge associated with the sciences proliferated widely during this period, mobilized for diverse intellectual, political, professional, and commercial purposes.[61] In the forensic field, experts in legal medicine such as Xu Songming

---

[61] Kwok, *Scientism in Chinese Thought, 1900–1950*; Schmalzer, *The People's Peking Man*, 17–54; Eugenia Lean, "Proofreading Science: Editing and Experimentation in Manuals by a 1930s Industrialist," in *Science and Technology in Modern China, 1880s–1940s*, ed. Jing Tsu and Benjamin A. Elman (Leiden: Brill, 2014).

never obtained a monopoly over the ways in which legal professionals utilized scientific knowledge. In this case, legal professionals themselves made significant claims over the physical evidence, in the process inter- preting scientific evidence in accordance with their own interests. Likewise, as we will see in Chapter 7, judicial officials might utilize science in ways that were not sanctioned by the experts – for example, requesting that medico-legal experts run laboratory tests on silver testing probes that had been used to confirm fatal poisoning in accordance with instructions in the *Washing Away of Wrongs*. This was not the world of "common knowledge" (*changshi*) that was disseminated by medico-legal professionals in order to discipline the evidence norms of judicial officials; rather, it was science that had been appropriated by judicial officials for their own needs and was, at times, only loosely congruent with the interests of professional scientists themselves.

Turning now to a controversy that unfolded on the pages of the news- paper *Shenbao* around the time of this case, we will examine a major challenge that physicians faced as they asserted their professional author- ity in homicide investigation: gaining legal and physical control over the dead body itself. The procedure of autopsy – opening the body and inspecting the internal organs to identify cause of death – was the forensic technique that most distinguished physicians' approach to the body from that of judicial officials and their body examiners. The claim of medical reformers that autopsying the body was the only way to reveal the ana- tomic-pathological changes that had led from life to death was one of the most powerful arguments that they had when attempting to convince judicial authorities to use their expertise in forensic cases. In practice, however, the challenges that physicians faced when they tried to extract bodies from the existing claims of police and judicial officials, not to mention those of relatives of the deceased, raised questions not only about the moral implications of physically damaging the dead body, but also about the necessity of their expertise vis-à-vis the already accepted examination practices of the judiciary.

6   Dissection and Its Discontents

Judicial officials in Shanghai, much like those in Beijing, demonstrated a keen interest in finding outside experts who could assist them in homicide investigation. During the early 1920s, the head of the local procuracy in Shanghai, Che Qingyun, consulted Dr. F. Oppenheim, a German pathologist and faculty member of Tongji University Medical School, about several difficult cases that he had encountered.[1] Pleased with the results, in late July 1924 Che entered a one-year contract with Tongji and the affiliated Paulun Hospital for assistance in forensic examinations. The procuracy's motivation, as it was recounted by Tongji affiliates later on, was to bring Chinese forensic practices up to the standards of other countries by having physicians conduct autopsies in its investigations rather than simply relying on its own forensic personnel to examine the body. Che viewed Chinese courts' use of this irrefutably modern medical procedure as advancing the cause of judicial reform and strengthening the case for foreign countries to give up their extraterritorial judicial privileges – a particularly pressing issue in Shanghai, a city in which foreigners still maintained extraordinary legal authority and jurisdiction in the Mixed Courts.[2] During the initial period of the agreement, as many as 300–400 corpses were reported for examination by medical school personnel. Of these, only about 50 were autopsied. An autopsy was only performed if "external inspection could not judge cause of death."[3]

---

[1] For Oppenheim's involvement with Shanghai judicial authorities, see SNB August 9, 1925, 15, as well as Yang Shangheng, "Jiying tichang zhi shiti pouyan" [On the urgent necessity of promoting the autopsy of corpses], *Tongji zazhi* no. 7 (1922): 24–5.

[2] For the impact of Shanghai's complex jurisdictions on Chinese lawyers' efforts to build a modern legal profession and the role of professionals in pressing for the rendition of the Mixed Courts, see Xu, *Chinese Professionals and the Republican State*, 215–41.

[3] Under this arrangement, the medical school made available a physician to assist judicial authorities in the forensic investigation of deaths. The procuracy provided the medical school with a monthly payment of 200 yuan which, affiliates of the school claimed, was just enough to cover transportation expenditures. The forensic examiners' salaries and other expenses associated with testing were paid by the school. SNB August 9, 1925, 15; SNB August 10, 1925, 15; Yang Yuanji, "Fayixue shilüe bu" [Supplement to *A Brief History of Legal Medicine*], *Beiping yikan* 4, no. 9 (1936): 11–12.

In mid-July 1925, a lawyer named Chen Kuitang presented the procuracy with a written petition requesting that the agreement not be renewed at the end of the month, the point at which it was set to expire.[4] Chen argued in his petition that centuries of use had proven the effectiveness of the *Washing Away of Wrongs* and that the "depth of the people's trust in it was apparent." Thus, it was unwarranted to replace these older methods with autopsy, a forensic technique that was not more effective and, moreover, caused physical damage to the corpse. According to Chen, the image of bodies broken open and organs removed was one that shocked the local people who were supposed to be served by this arrangement. As a result, people who would otherwise have reported cases of murder to officials for investigation now "resigned themselves to accepting an injustice but not being willing to seek its redress." For these reasons, Chen argued, the local procuracy should not renew the agreement:

In sum, a forensic examination [using the *Washing Away of Wrongs*] and dissection are both used to bring to light the injustices of the dead. If one uses dissection and the injustice is brought to light, and one uses an [inspection clerk's] examination and it is also brought to light, then it is the same result. Yet, if for one the corpse is ruined, and for the other it is maintained whole, how can those who govern the people only consider the redressing of wrongs and not also consider the corpse?[5]

Chen's petition was delivered to various public associations in the city and published in *Shenbao*, where it soon incited several responses.[6] In early August, the well-known medical reformer and ardent critic of Chinese medicine Yu Yunxiu published a letter in *Shenbao* on behalf of the Medical and Pharmaceutical Association of the Chinese Republic, criticizing the procuracy for entering into such an agreement in the first place.[7] The problem, Yu argued, was not in the procuracy's use of autopsy, but in the fact that the authorities had made such an arrangement with "Westerners" (*xiren*) rather than with Chinese forensic experts. During a period in which many in Shanghai's professional worlds of law and medicine were acutely concerned about the rendition of the Mixed Courts and Chinese claims of jurisdiction both within the city and nationally, it was unseemly for the procuracy to rely on German expertise for a task that was so essential for demonstrating the integrity of China's judicial system. The real issue, as Yu explained, was that given Europeans' propensity to maximize their advantage in any treaty into which they entered, affiliates of Tongji would use the provisions of the

---

[4] SNB July 16, 1925, 16.   [5] Ibid.   [6] SNB July 20, 1925, 15.
[7] SNB August 3, 1925, 14; Yu Yan, *Yu Yunxiu zhongyi yanjiu yu pipan* [Yu Yunxiu's research and criticisms regarding Chinese medicine], ed. Zu Shuxian (Hefei: Anhui daxue chubanshe, 2006), 369–70.

agreement to wantonly conduct as many autopsies as they wanted, even in cases for which the procedure was unnecessary, and in disregard of societal opinion. Given that the *Washing Away of Wrongs* was itself filled with anatomical errors and unsuitable for the modern age, as Yu demonstrated in a detailed critique of passages in the text, the only solution was to find Chinese physicians to handle the procuracy's forensic examination work.

A little less than a week later, affiliates of Tongji published their own response in *Shenbao*, which denied the veracity of Chen's claims. Not only were a small proportion of bodies actually autopsied under the agreement, the response noted, but, after such a procedure was carried out, "the body was sewn up and cleaned, just as if one had undergone surgery in a hospital. It is so that when the relatives see the body, they do not feel as if it had been physically deformed or damaged."[8] Beyond this, they did not address the underlying issues that Chen had raised – namely, the legitimacy of physically damaging the dead body when less invasive techniques could be used, and the implications of such violence on an "object" that was culturally meaningful. In the end, the procuracy revealed that it would not renew its contract with Tongji due to a lack of funding with which to meet the obligations of the agreement. Yang Yuanji, who had handled some of the forensic work on Tongji's side, claimed, rather, that the arrangement was discontinued because of the public uproar following Chen's petition.[9] In fact, by early August 1925 it was reported in *Shenbao* that the procuracy was already pursuing a similar arrangement with Gu Nanqun, a Japanese-trained physician who had founded Nanyang Medical College (Nanyang yixue zhuanmen xuexiao) and its affiliated hospital, and who was serving as superintendent of the hospital at the time.[10]

In studying the history of forensic autopsy in early twentieth-century China, historians have tended to side with Oppenheim and other forensic reformers in associating the use of anatomical dissection with modern progress and identifying the body examination practices of the *Washing Away of Wrongs* as a form of pre-modern forensic science simply waiting to be replaced with modern techniques. The historian of legal medicine Jia Jingtao, for example, has called the legalization of dissection in China in 1913 a crucial turning point in the advance from the forms of "ancient legal medicine" (*gudai fayixue*) used in imperial China to the "modern

---

[8] SNB August 9, 1925, 15.   [9] Yang, "Fayixue shilüe bu," 12.

[10] SNB August 5, 1925, 15. For more on Gu, see Li Yuanxin (William Yinson Lee), *Huanqiu Zhongguo mingren zhuanlüe: Shanghai gongshang gejie zhi bu* [World Chinese Biographies: Shanghai Commercial and Professional Edition] (Shanghai: Globe Publishing Company, 1944), 112–3.

legal medicine" (*xiandai fayixue*) that only started to be introduced from other countries during the late Qing and Republican periods.[11] From this perspective, which was held by Chen Kuitang's critics, resisting the use of autopsy was tantamount to denying China access to an essential, irrefutably superior technique of forensic investigation. More damningly, as Yu Yunxiu and others suggested in their critical discussions of the *Washing Away of Wrongs*, doing so implied accepting a body of forensic knowledge that was filled with easily recognized errors.

In fact, the impact of forensic autopsy on the investigation of homicide in 1920s and 1930s China is a far more complex question than such a singular focus on its purported epistemological superiority would suggest. As David Luesink has shown, dissection-based anatomical inquiry was essential for physicians' attempts to lay foundations for modern medical education in Republican China and to establish intellectual authority over the body – both crucial preconditions for building their profession.[12] At the same time, when one examines how physicians actually asserted their new claims over the dead body vis-à-vis those of others – including relatives of the deceased, journalists, police, and judicial officials – a deeply ambiguous picture of medical authority emerges. As essential as autopsy might have seemed to proponents of forensic reform, Chen's point that the procedure was redundant was, in fact, one that was often assumed in practice by judicial officials themselves. In this context, anatomical dissection carried complex and, at times, incongruous meanings, at once central to the assertion of professional medical authority, ambivalent in public opinion, or simply unnecessary for Chinese judicial officials who already made their own recognized claims over the dead body.

Autopsy and Its Alternatives

It is clear from the response that Tongji affiliates published in *Shenbao* that they did not share Chen Kuitang's concerns with the moral or cultural implications of dissecting cadavers. The focus of their response was another issue: Chen's claim that dissection-based forensic examination practices were equal in effectiveness to those of the *Washing Away of Wrongs*. In a lengthy critique of the text published in the newspaper, Oppenheim and an assistant instructor in pathology named Du Keming critically evaluated various passages according to modern medical understandings of

---

[11] Jia Jingtao, "Xinhai geming yihou de zhongguo fayixue" [Legal medicine in China after the revolution of 1911], *Zhonghua yishi zazhi* 16, no. 4 (1986): 205.
[12] Luesink, "Dissecting Modernity."

anatomical structure and pathological change. For example, in addressing the long-standing method of using knowledge of the body's vulnerable points (the so-called "vital spots") to decide which of several wounds had caused death, Oppenheim explained that a wound's location on a point on the surface of the body designated as a "vital spot" was not in itself a sufficient criterion for deciding that it was the mortal wound:

> If a wound becomes infected or too much blood is lost, then even if it is on a non-vital spot it can still cause death. If a wound is only skin deep, then even if it is on a vital spot, it will not cause death. In deciding if a particular wound has caused the death, one must examine internal injuries, whether or not an injured organ is vital, and the severity of the wound. Moreover, some severe wounds that cause death are internal leaving no mark on the outside of the body. The *Washing Away of Wrongs* makes little mention of examining the inside of the body. This is the greatest flaw of the book.[13]

Underlying such critiques of judicial officials' forensic practices was the assumption that there was only one source of valid knowledge of the body, based on Western scientific medicine, and that any examination practice that failed to meet this epistemological standard was categorically unable to attain the same level of certainty and authority. As the growing literature of medical reformers' critical evaluations of the *Washing Away of Wrongs* showed, homicide cases in China were routinely decided on the basis of ideas about the body that were easily disproven through the application of basic knowledge of anatomy, or, in the case of Lin Ji's critical research on the silver needle test, simple chemistry experiments. This strategy of claiming professional authority on the basis of their exclusive, specialized knowledge of the physical world was fundamental to forensic reformers' attempts to reorganize forensic examination as an area of professional medical expertise, much as it was for critics of Chinese medicine who used similar strategies to challenge Chinese-style physicians' understandings of physiology or pharmacology.[14] In promoting the use of autopsy, forensic reformers such as Oppenheim were making the argument that it was only by carefully inspecting the condition of the internal organs that one could accurately determine what had caused death – a precondition for deciding questions of legal culpability. Without autopsying the body, proponents of forensic reform argued, it was impossible to establish an accurate and complete diagnosis of cause of death and to ensure that no pathological signs went unaccounted for. To illustrate this point, Sun Kuifang

---

[13] SNB August 11, 1925, 15.
[14] Luesink, "State Power, Governmentality, and the (Mis)remembrance of Chinese Medicine."

published a series of cases in *Monthly Bulletin of Legal Medicine* demonstrating that bodies with minor external wounds could harbor significant internal injuries.[15]

There was thus a clear disjuncture between the epistemological assumptions of Chen Kuitang and those of his critics, who denied the equivalence Chen claimed between the effectiveness of the *Washing Away of Wrongs* and that of dissection-based forensic examination practices. In fact, in questioning the assumed epistemological superiority of the latter, Chen was raising questions that were rarely articulated yet were of essential importance to China's forensic scene in the 1920s and 1930s: Why should judicial officials adopt a new body examination technique when they already had techniques that were acknowledged to be effective in practice and that did not usually require the opening of the corpse? What was it about autopsy that justified the use of such a procedure?

It is important to note from the outset that there was no law requiring that judicial officials seek the assistance of physicians in their cases. The promulgation of a legal framework that permitted and regulated medical dissection in November 1913 was a watershed moment in the history of forensics in twentieth-century China, much as it was for Western scientific medicine more generally. China's anatomy laws defined the circumstances under which bodies could be dissected, the classes of unclaimed cadavers that could be lawfully obtained by medical schools, and the relationships between physicians, relatives of the deceased, and officials surrounding the dead body. The original five-article law was supplemented in April 1914 by regulations that added further detail to claiming and reporting procedures, and two more iterations of these laws were promulgated over the next two decades.[16] From the earliest set of regulations onward, these laws established that a physician could be called by judicial authorities to conduct an autopsy in cases of unnatural death when cause of death could not be ascertained otherwise. This was, in fact, a fairly narrow construal of a physician's role in policing and the administration of justice, especially when compared to the expansive regulations on physicians' forensic involvement that were

---

[15] Sun Kuifang, "Zaoqi pouyan zhi zhongyao" [The importance of early autopsy], *Fayi yuekan*, no. 17 (1935): 1–10.

[16] For the role that Tang Erhe played in the establishment of these laws and their larger significance for medical professionalization and modern Chinese governance, see Luesink, "Dissecting Modernity," 236–42. For the 1913 regulations and 1914 supplementary regulations, see Zhang and Xian, *Minguo yiyao weisheng fagui xuanbian, 1912–1948*, 1–3. A new set of regulations was promulgated by the newly established Nationalist government in May 1928 and revised in June 1933. For the 1928 regulations, see *Zhonghua yixue zazhi* 14, no. 3 (1928): 203–4. For the 1933 revised regulations, see Zhang and Xian, *Minguo yiyao weisheng fagui xuanbian, 1912–1948*, 162–4.

being promulgated in other countries. Russia and Sweden, for example, had established detailed regulations defining the nature of the physician's role in such cases, even specifying the examination and reporting protocols that had to be followed.[17] The closest that Chinese law came to these kinds of provisions was, ironically perhaps, the official form that defined the procedures and terminology that inspection clerks had to use in their examinations.

Despite the fact that physicians lacked a legal framework that would compel judicial officials to accept their participation in such cases, the use of autopsy did, in certain instances and under certain conditions, attain authority in Chinese law during this period. This fact alone should not be surprising. As David Luesink has shown, anatomy was a field of academic and practical inquiry that purported to provide scientific solutions to some of the most pressing questions of state power and racial and national identity.[18] The medicine of anatomic pathology – a basic foundation on which the forensic approaches of Oppenheim and other proponents of legal medicine were based – gained authority in China within the larger context of Chinese responses to Euro-American military and political domination and the establishment of new state institutions and bodies of expert knowledge that were necessary to demonstrate the country's modernity. Broadly speaking, these were the same political conditions in which hygienic modernity (*weisheng*) became an imperative of Chinese governance during this period, and one that raised similarly pressing questions of political sovereignty.[19]

Despite the scientific cachet of the technique, when we look at the ways in which forensic autopsy was actually used in cases, we find that it provided judicial officials with a new way to answer the same basic questions that had always pertained to homicide investigation: Had a death been caused by the actions of another person (homicide)? Or was it a case of suicide or other circumstances that absolved others of legal responsibility? Officials in China had long relied on subordinate body examiners to answer these questions and, from the perspective of procurators, there was little that was new about the *function* that autopsy played in homicide investigation. As an example of the ways in which autopsy could be used in such cases, we might examine the case of an unidentified deceased male that was investigated by personnel of the Research Institute of Legal Medicine in August 1933. The case began with the discovery of a body in a Chinese-administered area of Shanghai

---

[17] Becker, *Medicine, Law, and the State in Imperial Russia*, 199–209; Eva Åhrén, *Death, Modernity, and the Body: Sweden 1870–1940* (Rochester: University of Rochester Press, 2009), 31–2.
[18] Luesink, "Dissecting Modernity."   [19] Rogaski, *Hygienic Modernity*.

located just outside of the northeastern-most corner of the International Settlement.[20] Much as in the case of other corpses found in the city, it was the local headman (*dibao*) of the area who initially handled the case and accepted primary responsibility for burying the body later on.[21] Having arrived at the scene, this individual inspected the clothing and other items found on the body and examined the body itself. Because this preliminary examination yielded that the "cause of death was unclear," the headman notified the Shanghai procurator's office to initiate an official investigation.

Rather than using its own forensic personnel to examine the body, this office contacted the Research Institute of Legal Medicine and entrusted its experts with the case. This decision reflected the significant role that this institution had come to play in criminal investigation in Shanghai since its establishment in late 1932. Beginning in March 1933, the Research Institute began to handle "all of the regular cases" (*yiqie putong anjian*) of the Shanghai Local Court, which amounted to about 140 to 150 cases per month. These included the inspection of wounds on the living, the testing of drug offenders, inspections of corpses, and a small number of other cases including investigations of rape, illness, and appraisals of virginity. By July 1933, the Research Institute had handled 2,200 such cases in total; by August 1934, its personnel had handled over 6,600.[22] There is a significant contrast here with the situation in 1930s Beiping, as we have seen, where the procuracy and its examiners maintained a firm hold over the routine investigation of unexpected deaths and homicides in the city. Physicians of Western scientific medicine never made comparable inroads in the routine investigation of deaths in Beiping despite the fact that this city too would become an important regional center for new laboratory-based forensic testing services during the 1930s.

After taking possession of the corpse in this case, personnel of the Research Institute – almost certainly the hands-on Lin Ji, who was director at the time – conducted a thorough examination of the body that was reported in minute detail in the written appraisal report that was returned

[20] The appraisal report from this case was published in *Fayi yuekan*, no. 9 (1934): 145–9.
[21] For more on the role that local headmen played in the collection and burial of abandoned bodies in the International Settlement, see Henriot, "Invisible Deaths, Silent Deaths," 421–2.
[22] Lin Ji, "Sifa xingzheng bu fayi yanjiusuo chengli yizhounian gongzuo baogao" [Report on the work carried out by the Research Institute of Legal Medicine of the Ministry of Judicial Administration in the year since its founding], *Fayi yuekan*, no. 1 (1934): 6; "Fayi yanjiusuo juxing erzhou jinian Lin suozhang baogao jingguo" [The Research Institute of Legal Medicine observes its second anniversary, Director Lin reports on its progress], *Zhonghua yixue zazhi* 20, no. 8 (1934): 1101.

to the procuracy. The autopsy proceeded from an inspection of the exterior surfaces of the body to the internal organs to an examination of tissue specimens taken from each. Lin Ji concluded on the basis of widespread hemorrhaging and engorgement of blood throughout the body that the death had been caused by heat stroke, with nephritis and fatty heart as secondary factors that could have contributed to the man's death. Lin Ji also made a number of observations in the report that served to rule out homicide, including the fact that there were no wounds on the exterior of the body and no signs of poisoning in the gastrointestinal tract. The written appraisal concluded with an explanation of the anatomic-pathological signs and causes of heat stroke, a discussion meant to provide judicial officials with enough basic medical knowledge to understand the significance of the autopsy findings.[23] It is likely that inspection clerks would have come to the same conclusion that the man had died of illness, not homicide or suicide, even though they would have followed a different reasoning process in doing so.

For procurators in 1920s and 1930s China, the most important question in such cases was whether or not anyone should be held criminally responsible for the death. The main goal was to identify the "manner of death" – that is, the categorization of deaths as due to homicide, suicide, accident, or natural causes, classifications that pertain to the social and legal implications of a death as much as to the bodily, or medical, cause. Forensic death investigation was, for these officials, an activity geared almost exclusively toward legal questions. As such, the careful accounting of the body that experts like Lin Ji could offer in cases such as this one contained medical details that, while demonstrating thoroughness, were not immediately relevant to the legal questions at hand. As we have seen in earlier chapters, claims about manner of death were routinely made in Beijing on the basis of witness and police statements, an examination of the scene, and an inspection clerk's confirmation of the wounds (or lack thereof) found on the surface of the body. In claiming that autopsy was indispensable to homicide investigation, forensic reformers were arguing that judicial authorities' commonsensical approaches to such cases were inadequate given their inability to examine anatomic-pathological signs that were only observable under the skin.

It was undoubtedly easy for his critics to cast Chen Kuitang as an individual whose opinions were out of touch in an age in which rapid advancements in forensic science had rendered older forms of knowledge obsolete. While this kind of explicit public defense of the *Washing Away of Wrongs* was rare in 1920s and 1930s China, it is clear that many, including

[23] *Fayi yuekan*, no. 9 (1934): 149.

judges and procurators, accepted the legitimacy of using the text to guide body examinations. The print media routinely depicted inspection clerks as legitimate participants in the everyday administration of justice and policing, coverage that also elevated their claims about the body to the status of journalistically validated facts. In fact, the equivalence assumed by Chen Kuitang between the effectiveness of the *Washing Away of Wrongs* and forensic procedures based on Western scientific medicine was not dissimilar to the forensic pluralism of judicial officials, who routinely demonstrated flexibility and pragmatism when deciding when and how to use and combine these methods in their cases. Viewed from this perspective, Chen Kuitang's claim that medical autopsy was not indispensable to the administration of justice was less an irrational response to modern progress, as his detractors characterized it, than an assessment of the strengths and weaknesses of new forensic techniques vis-à-vis old ones, as well as a critique of a new professional group's claims over the dead body. Turning now to another controversy surrounding the use of autopsy that unfolded in Beiping in the early 1930s, we will see that Chen Kuitang was not the only critic of the new relationships between physicians, relatives of the dead, and the state that were forged around anatomical dissection.

The Case of Song Minghui

It is worth drawing a distinction between the intellectual authority of dissection-based anatomical knowledge in 1920s and 1930s China and the ambivalent cultural and social meanings that might be attached to the actual practice of claiming and dissecting corpses. Anatomy maintained an important status within many areas of the early twentieth-century sciences, including clinical medicine, pathology, and physical anthropology.[24] Beyond the significance of anatomy to the sciences themselves, as Ari Larissa Heinrich has argued, Chinese-language anatomical works such as Benjamin Hobson's *A New Treatise on Anatomy* (Quanti xinlun, 1851) also introduced a new "aesthetics of anatomical realism" to the nineteenth- and early twentieth-century Chinese intellectual and cultural scene.[25] It was on the basis of these new ways of understanding the interiority and layered materiality of the body as well as new assumptions about empirical observation as a source of objective bodily knowledge that later authors of medical and literary works, including Lu Xun, made new kinds of aesthetic claims to the real. The fact that judicial officials themselves drew on anatomical

---

[24] Luesink, "Dissecting Modernity," 211–47.
[25] Larissa Heinrich, *The Afterlife of Images: Translating the Pathological Body between China and the West* (Durham: Duke University Press, 2008), 113–47.

knowledge contained in one of the early translations of Japanese legal medicine to revise the forensic examination forms that were used by the judiciary's forensic personnel, one of a number of instances in which officials drew on medico-legal science to supplement the existing forensic system, suggests the status that Western and Japanese anatomy had already attained as a form of authoritative knowledge about the body.[26]

When it came to actual dead bodies, to the procedures through which they were claimed by physicians and the social politics surrounding this process, the question of what anatomical dissection signified was less clear-cut. Claiming bodies for dissection could be contentious and controversial given the frictions that could result when physicians' new claims over the dead encountered those of the others who were already invested in establishing the meanings of the dead body. We might examine a controversy that occurred several years after the publication of Chen Kuitang's petition to explore the ambivalent meanings that could emerge surrounding physicians' claims over the dead body. The focus of this case was the autopsy of Song Minghui, a 34 *sui* male who was admitted to the hospital of PUMC in late July 1930 following a disastrous encounter with therapeutic cauterization that left severe burns on his legs.[27] Upon Song's death on July 31, the hospital notified both his family and Beiping judicial authorities, who immediately carried out a forensic inspection of Song's corpse in order to collect evidence for their legal inquiry into the accident that had preceded his hospitalization. Following this examination, Song's body was autopsied by Ai Shiguang (Ngai Shih-kuang), a recent graduate of Xiangya Medical College and Assistant in Pathology at PUMC who routinely handled hospital autopsies.[28] It is clear from the findings of Ai's post-mortem examination that Song entered the hospital with a number of other medical conditions, including calcified areas of the lungs and lymph nodes that would suggest chronic tuberculosis.[29]

---

[26] These forms were revised on the basis of Xu Lian's *Xiyuan lu xiangyi* (1854) and the anatomical images of a text that officials of the Ministry of Justice referred to as *Shiyong fayixue*, seemingly Wang You and Yang Hongtong's 1908 translation of Ishikawa Kiyotada's *Practical Legal Medicine* (Jitsuyō hōigaku). See "Banfa yanduanshu jianduanshu bing shangdan geshi ling," 250 as well as Wang You and Yang Hongtong, *Shiyong fayixue daquan*, 40–1, which contains one of the anatomical diagrams on which officials of the Ministry drew in revising the form.

[27] This account is derived from newspaper coverage, including transcripts of judicial authorities' interrogation of Song Minghui's family members and PUMC physicians. SJRB August 12, 1930, 7; SJRB August 21, 1930, 7; SJRB October 2, 1930, 7.

[28] For Ai's appointment at the time, see Minutes of the Peking Union Medical College, Medical Faculty Executive Committee, February 28, 1930, 2930–31, CMB, Inc., Box 8, Folder 52, RAC.

[29] For a record of the autopsy, see "Anatomical diagnoses, autopsy nos. 601–1000," autopsy no. 962, CMB, Inc., Box 10, Folder 67, RAC.

Song's widow, née Cui, and brother Song Yueshan soon arrived at the hospital with the stated intention of collecting the body for burial. According to the harrowing account given in statements of the Song family that were later printed in the newspaper *World Daily*, hospital staff initially did not permit them access to the body. It was only when they disregarded these instructions and went further into the building that they found Song's corpse, which, it was now clear, had been opened during an autopsy. According to the statement of Song's widow, who recounted the scene, "I then asked them 'why did you damage my husband's body?' They said it was 'to study the disease.' I then asked again: 'He has been dead for two days and two nights and the authorities already inspected the corpse. How can it be that you still need to study the disease?'"[30] Following this exchange with hospital staff, Mrs. Song née Cui fainted, and other members of Song's family tried to carry her out of the hospital, only to be blocked by guards at the exit. When hospital staff suggested giving her medicine, Song Yueshan reportedly quipped "Don't take the medicine! Will another person take your medicine and then die as well?"

Song's family subsequently brought the case to the Beiping procurator's office, accusing Ai Shiguang and Wang Ligeng (Wang Li-keng), an Assistant Resident with whom Song's family had consulted, of dissecting Song's body illegally without having obtained their permission.[31] Since the first regulations on dissection promulgated in November 1913 and continuing into the early 1930s, physicians had been legally required to obtain relatives' permission before conducting an autopsy, even in cases for which an individual had given consent for the procedure while alive.[32] According to Wang Ligeng, Song Yueshan had in fact given his permission for the autopsy and signed the form with his right thumbprint. Song denied signing the permission form, claiming that he had only given his thumbprint on other paperwork, which had been required upon Song Minghui's admission to the hospital. The autopsy permission form, Song's family now argued, had been fabricated by hospital staff who had attempted to trick the illiterate Song Yueshan into allowing his brother's body to undergo the procedure.[33]

It is likely that the Song family had faced some degree of pressure from medical staff to permit the autopsy to go ahead. Over the 1920s and

[30] SJRB August 12, 1930, 7.

[31] For Wang's appointment, see Minutes of the Peking Union Medical College, Committee on the Hospital, March 4, 1930, 2930–45, CMB, Inc., Box 5, Folder 32, RAC.

[32] For the "Regulations on the dissection of corpses" in effect at the time, see *Zhonghua yixue zazhi* 14, no. 3 (1928): 203.

[33] For these conflicting accounts, see SJRB August 12, 1930, 7; SJRB August 24, 1930, 7; SJRB September 3, 1930, 7; SJRB October 2, 1930, 7.

1930s, the faculty and staff of PUMC pursued various strategies to address the "very difficult and often disagreeable task," in the words of one member of the school's administration, of persuading relatives to give their permission for autopsy.[34] One strategy pursued in late 1924, for example, was to have medical superintendent of the hospital Wang Xichi (S.T. Wang, 1893–?) personally attend to relatives of patients whose death was imminent and "[take] pains to become acquainted with the members of the family and to do special things for their comfort," thus creating a foundation of good will for later discussions of autopsy.[35] The importance of such "diplomacy and tactfulness" was codified in the general manual of hospital policies and procedures that was in use during the early 1930s:

Success in securing permission for post-mortem examination rests chiefly on the efforts of the assistant resident and interne. Their intimate relation to the patient and his family should build up confidence and understanding to make possible an easy approach. Tact is essential. The purpose should be clear and misunderstandings should be carefully avoided. The method used must vary with the social status and intelligence of those concerned. Even though permission is not secured, the family should be impressed with the good intentions of the hospital and with the fact that everything possible has been done for the patient.[36]

It is likely that these negotiations were less straightforward in practice. Throughout the late 1920s and early 1930s, ongoing discussions were taking place among PUMC administrators and faculty about instituting a requirement that relatives sign the autopsy permission as a condition of a patient's admission to the hospital.[37] In early November 1931, officers of the school approved changes in general hospital regulations requiring that relatives of foreign patients give such a written expression of consent for autopsy upon admission, a policy that, as Roger S. Greene noted, might "set a valuable precedent which may later affect the practice in respect to Chinese patients."[38] The fact that it was Wang Ligeng, the hospital staff member most directly responsible for securing the

---

[34] Harriet Barchet to Margery K. Eggleston, December 23, 1924, CMB, Inc., Box 68, Folder 482, RAC.

[35] Ibid.

[36] O.H. Robertson and S.N. Cheer, *Manual for the Medical Services of the Peiping Union Medical College Hospital*, 3rd edition. (Peiping: PUMC Press, 1930), CMB, Inc., Box 69, Folder 487, RAC.

[37] Roger S. Greene interview with J.R. Cash, August 6, 1927, CMB, Inc., Box 114, Folder 828, RAC; Roger S. Greene interview with T.F. Huang, May 27, 1929, CMB, Inc., Box 68, Folder 482, RAC.

[38] Minutes of the Peiping Union Medical College, Committee of the Hospital, November 9, 1931, 3132–31, CMB, Inc., Box 5, Folder 34, RAC; Roger S. Greene to Margery K. Eggleston, January 16, 1932, CMB, Inc., Box 68, Folder 482, RAC.

permission, who signed Song Minghui's autopsy permission form as a witness (*jianzheng ren*) suggests, at the least, the hospital's toleration of a conflict of interest produced under the institutional pressure to carry out autopsies.[39]

In the end, investigating authorities found no evidence of wrongdoing on the part of the hospital in this case.[40] The Song family subsequently appealed to the procuracy of the High Court and, following its decision to not pursue the case, lodged an appeal with the Supreme Court.[41] Meanwhile, as the legal maneuvers of the Song family and their unnamed supporters were moving forward, a bizarre chapter in the case unfolded in early September 1930 with the news that a bomb had exploded at PUMC, an incident that was immediately interpreted by some as the result of the enmity generated by the autopsy of Song Minghui.[42] The bomb had been placed in a suitcase and left in a hospital waiting area several days earlier. It exploded, reportedly, when a member of the administrative staff attempted to open the bag in order to identify its owner.

A no less explosive battle was being waged on the pages of the north China press, which was covering the case in terms that were highly unfavorable to PUMC. In late August 1930, readers of *World Daily* were informed that the Beiping Local Court had received a flurry of letters attacking PUMC with the outlandish claim that Song Minghui's fate was one example of a more pernicious pattern – namely, that "when the hospital discovers that patients have a particularly unusual disease, they use drugs to kill them and cut up the bodies for their research."[43] Thus, the newspaper reported, some "city people" (*shimin*) were now urging the court to investigate the autopsy of Song Minghui's body as not only an illegal autopsy, but also as a possible case of homicide as well. Readers of *World Daily* were also exposed to ongoing criticism of PUMC in the commentary of Wang Zhuyu, a well-known reporter in the city whose daily column addressed news items of the day. Following the procuracy's decision to not go forward with the case, for example, Wang noted that

---

[39] SJRB August 24, 1930, 7.   [40] Ibid.

[41] SJRB September 6, 1930, 7; SJRB October 2, 1930, 7; SJRB January 23, 1931, 7.

[42] For an account of the immediate circumstances surrounding the explosion, see SJRB September 5, 1930, 7. The story was picked up by the Associated Press, which reported that "[the] police believe that the affair has some connection with a recent Chinese legal case. Relatives of a Chinese patient who died in the hospital unsuccessfully sued two doctors of the college, charging that they had performed an autopsy on the body without consent of the family." Also see brief mention of the incident in Margery K. Eggleston to Paul Hodges, September 8, 1930, CMB, Inc., Box 125, Folder 905, RAC. A copy of *New York Sun* (September 4, 1930) coverage of the incident, which contains the A.P. report, is included in this file.

[43] SJRB August 22, 1930, 7.

despite judicial authorities' evidence to the contrary, it was highly unlikely that an "ignorant fool" such as Song Yueshan would ever knowingly provide permission for his brother's body to undergo an autopsy.[44] Moreover, while the procuracy had claimed that it was unreasonable to think that hospital staff would take Song Minghui's fingerprint with the intention of forging an autopsy permission later on, the "idea that the hospital would take a fingerprint in advance" for this purpose was, according to Wang's own information, accurately describing "a common occurrence."

Given the tenor of this reporting, supporters of PUMC were quick to direct criticism toward journalists for their prejudicial tone and for failing to check their facts before reporting developments in the case. An important source of such criticisms was the Bingyin Medical Society (Bingyin yixue she), an organization that had been established in 1926 by a group of self-avowedly patriotic PUMC students, including the well-known medical reformer and proponent of rural healthcare Chen Zhiqian (C.C. Chen), for the purposes of promoting popular knowledge of Western scientific medicine and hygiene.[45] One of the most important activities that the members of this association carried out was the publication of newspaper supplements containing popularized medical knowledge and opinion pieces on the state of medical affairs in China. In early July 1929, for example, members of the Society began to publish a semi-monthly medical feature in *L'Impartial* (Dagong bao) that became a weekly supplement in the newspaper in November of the same year.[46] It was in this venue, and in widely read collected volumes of the society's articles, that members of this group attempted to defend the reputation of PUMC and the legitimacy of Western scientific medicine while holding journalists accountable for their treatment of the case.[47]

One of the most active of such commentators was Jia Kui, a member of the PUMC graduating class of 1926 and one of the founding members of the Bingyin Medical Society.[48] Over the month of August 1930, Jia published several pieces in the society's weekly supplement in *L'Impartial*

---

[44] SJRB August 24, 1930, 7.

[45] Hu Yifeng, "Bingyin yixue she chutan – chengli beijing, zaoqi huodong yu lishi yiyi" [An initial study of the Bingyin Medical Society – Its background, early activities and historical significance], *Beijing dang'an shiliao* no. 3 (2005): 184–96.

[46] For an overview of the contents of this medical supplement, see Li Xiuyun, *"Dagongbao" zhuankan yanjiu* [Research on specialized supplements in "Dagongbao"] (Beijing: Xinhua chubanshe, 2007), 130–48.

[47] Jiang Shaoyuan, "Lianhe qilai yonghu poushi" [Unite to support the autopsying of cadavers], *Yixue zhoukan ji* 4 (1931): 268–70.

[48] For a brief biographical entry on Jia, which includes year of graduation, see Bullock, *An American Transplant*, 234.

criticizing journalists for their lack of "common knowledge" (*changshi*) of science and medicine, for disseminating misinformation about PUMC's involvement in the case, and, more generally, for failing to present anatomical dissection in a favorable light.[49] Jia argued that, if left unchecked, such coverage could damage the reputation of Western scientific medicine and fuel misguided public anxiety about dissection. For example, Jia criticized journalists of the newspaper *Jingbao* for making the erroneous claim that PUMC's autopsy permission forms were printed in a foreign language and for describing the condition of Song Minghui's dissected body as "awful" (*can*) in the title of one of its articles, a characterization that seemed to betray their prejudices against the procedure.[50] Given the power that journalists had over the direction of public opinion, Jia argued, their irresponsible reporting could have great implications for the public legitimacy and acceptance of Western scientific medicine in China.[51]

It is all too easy to interpret this controversy, as well as that surrounding Chen Kuitang's petition, as a reflection of Chinese society's pervasive conservatism when it came to anatomical dissection, an assumption held by some who commented on the progress of anatomical practice in China in the wake of these cases.[52] As Ari Larissa Heinrich has noted, it was a common trope in the writings of missionaries and others during this period that culturally ingrained attitudes about the body had hindered the progress of anatomical knowledge and dissection. From these observers emerged a conception of "the lack of willingness or 'ability' to perform autopsy because of what [Western medical missionaries] saw as the cultural superstition that prevented it."[53] Early twentieth-century proponents of dissection who were frustrated by the meager supply of corpses made available to them for medical training and research likewise argued that difficulties in procuring cadavers reflected the influence of indigenous "customs" (*fengsu*) that precluded a willingness to accept the postmortem dissection of one's own body or those of relatives.[54] Those who

[49] Youxian, "Wei yixue qingyuan yu xinwenjie" [A petition for the journalism profession on behalf of medicine], *Yixue zhoukan ji* 4 (1931): 253–5; Youxian, "Zhi Henshui xiansheng yi feng gongkai de xin" [An open letter to Mr. Henshui], *Yixue zhoukan ji* 4 (1931): 255–6; Youxian, "Xiang Jingbao jizhe jin yi yan" [A word to the reporters of *Jingbao*], *Yixue zhoukan ji* 4 (1931): 257–60.

[50] Youxian, "Xiang Jingbao jizhe jin yi yan," 259.

[51] Youxian, "Wei yixue qingyuan yu xinwenjie."

[52] Lu Runzhi, "Jiepou shiti guize zhi piping" [Criticisms of the *Regulations on the dissection of corpses*], *Yiyao pinglun* 5, no. 1 (1933): 11–4. For part two, see *Yiyao pinglun* 5, no. 2 (1933): 7–9; Li Tao, "Jiyi xiuzheng zhi jiepou shiti guize" [On the necessity of revisions to the *Regulations on the dissection of corpses*], *Zhonghua yixue zazhi* 16, no. 6 (1930): 529–34.

[53] Heinrich, *The Afterlife of Images*, 118–19, 134–5.

[54] Lin Zhen'gang, "Liunian jian bingli jiepou zhi tongji di guancha" [A statistical survey from six years of pathological dissections], *Guoli Beijing yixue zhuanmen xuexiao shizhou*

have sought to explain the perceived difficulties involved in getting the Chinese to accept dissection have commonly invoked lines from the *Classic of Filiality* (Xiaojing) that seem to exemplify the Chinese cultural aversion to the opening of the body: "Our body, skin, and hair are all received from our parents; we dare not injure them. This is the first priority in filial duty."[55]

Dissection could be controversial in early twentieth-century China, to be sure. In one sense, this fact alone should not be surprising. Schemes for collecting cadavers for medical purposes were controversial in nineteenth- and early twentieth-century Europe and the United States as well, at times inciting violent outbreaks when local people perceived that anatomists and allied grave robbers had transgressed contemporary codes of moral behavior surrounding the dead.[56] It would be more surprising if arrangements to claim the dead body for anatomy had not incited any negative response in China at all, especially given that the actions of medical schools could become implicated in questions of foreign exploitation, Euro-American imperialism, and cultural insensitivity. These associations were made, for example, in one of the Song family's later appeals, which accused PUMC of, among other things, "making the bodies of our countrymen into their test specimens (*shiyanpin*)."[57] The notion that Chinese instinctively opposed dissection on the grounds of custom or ritual, however, is a different claim, and one that ought to be located in relation to other essentializing discourses that grew out of modern attempts to explain China's purported deviation from seemingly universal – albeit in reality highly Eurocentric – models of Western science and cultural development.[58]

Indeed, while those who promoted anatomical dissection in early twentieth-century China often voiced the assumption that this procedure presented unprecedented challenges to existing Chinese death customs

jinian lunwenji (Beijing: National Medical College, 1922), 125; Li, "Jiyi xiuzheng zhi jiepou shiti guize," 530.

[55] "Shenti fafu shou zhi fumu, bu gan huishang, xiao zhi shi ye." Translation taken from Wm. Theodore de Bary and Irene Bloom, *Sources of Chinese Tradition: From Earliest Times to 1600*, 2nd edition. (New York: Columbia University Press, 1999), 326. For sources that cite this passage, see, for example, Zhu Hengbi, "Jiepou shiti zhi shangque" [A discussion on the dissection of corpses], *Zhonghua yixue zazhi* 8, no. 4 (1922): 198–207; Li, "Jiyi xiuzheng zhi jiepou shiti guize," 530.

[56] Ruth Richardson, *Death, Dissection, and the Destitute* (Chicago: The University of Chicago Press, 2000 [1987]), 90–3; Michael Sappol, *A Traffic of Dead Bodies: Anatomy and Embodied Social Identity in Nineteenth-Century America* (Princeton: Princeton University Press, 2002), 106. For public resistance to dissection in Sweden during this period, see Åhrén, *Death, Modernity, and the Body*, 17–49.

[57] SJRB August 21, 1930, 7.

[58] Michael Adas, *Machines as the Measure of Men: Science, Technology, and Ideologies of Western Dominance* (Ithaca: Cornell University Press, 1989).

and norms, questions surrounding the moral and legal implications of physically damaging the dead body were not as new as they seemed. Central government and local officials of the Qing and Republic and relatives of the dead had long had to negotiate the moral trade-offs and legal liabilities that arose when pursuing justice required physically harming the corpse – the very issues that emerged so centrally in Chen Kuitang's petition and in the case of Song Minghui. Skeletal examinations too could be viewed as physically damaging the corpse in "awful" (*can*) ways, an affective description that might be applied to the dissected body as well. Nonetheless, this concern neither stopped relatives from demanding the re-examination of remains in cases with disputed forensic evidence, nor did it keep judicial officials from carrying it out. That Song Minghui's body remained at PUMC, unclaimed by his family for months after the disputed autopsy, also suggests their deployment of a strategy that, as we have seen, was used in other cases in which relatives wanted to place pressure on another party in a dispute.[59] This was a very pragmatic use of the dead body, and one that would seem to run counter to the "customs" (*fengsu*) that, it has often been assumed, reflected Chinese veneration of the dead. In other words, even for the family of one who died, sometimes it was preferable to not put the dead to rest.

Policing Dissection in Beijing

For the physicians who were involved in China's early Western medical profession, conducting forensic autopsies in the service of judicial authorities was less of a priority than obtaining cadavers for the training of medical school students and academic research in anatomy and pathology. As in the case of Song Minghui, for example, deceased hospital patients could be autopsied for the purpose of investigating pathological conditions or changes that led to disease or death. Such a procedure could be carried out to confirm the earlier diagnosis of a physician, assess the consequences of a chosen course of clinical action, or, taken in aggregate, inform broader analyses within pathology, public health, or other fields of medical knowledge.[60] These autopsies, conducted by hospital or medical school personnel, were usually not used as evidence in legal proceedings. The primary goal was to solve medical problems, not legal ones. Assisting in homicide investigation was an arena in which physicians could assert professional authority, as we have seen. Yet, the other applications of

---

[59] As of late January 1931, Song Minghui's body remained unclaimed by his wife and brother, remaining in a refrigeration facility at PUMC. SJRB August 31, 1930, 7; SJRB January 19, 1931, 7.

[60] Lin, "Liunian jian bingli jiepou zhi tongji di guancha."

anatomical dissection were of relatively more importance to the production of academic and clinically relevant knowledge of disease and the body and the training of new physicians, essential areas of academic research and professional practice. In order to gain access to cadavers for these purposes, medical schools and hospitals had to enter into relationships with officials that could be quite different from those that applied when physicians served as outside experts in the service of a court. In such instances, rather than playing the role of experts to whom police or legal officials looked for assistance, physicians and their professional activities became objects of official supervision.

The Beijing police exercised significant oversight over hospital autopsies. The involvement of the police in such cases was sanctioned by the regulations on dissection promulgated in November 1913 and April 1914, which established that medical schools were required to notify local authorities prior to dissecting, even in cases for which relatives had already given permission for the procedure.[61] It is clear from autopsies conducted at PUMC that the Beijing police routinely insisted on investigating whether the proper permissions had been given, directly questioning relatives and instituting other measures to establish clear lines of responsibility surrounding hospital autopsies.[62] For example, we find a great degree of police oversight in the case of Ding Yulin, a patient of the old hospital of Union Medical College, the missionary-run medical school that was acquired by the China Medical Board of the Rockefeller Foundation in 1915 and subsequently expanded to form PUMC. Ding died in early September 1919 after unsuccessful treatment for liver disease, and police of the Inner Left First district soon became involved in establishing whether or not the body could be autopsied. Police officers soon questioned Ding's in-laws at the district police station, established that they "completely bore all responsibility" for permitting the procedure in lieu of the permission of Ding's parents in Hangzhou, and dispatched an officer to report on the autopsy and the disposal of Ding's body.[63]

In the discussions that were taking place between faculty and administrators of PUMC, such actions by the police became part of a broader narrative of official authorities' efforts to obstruct the carrying out of autopsies. Writing on the cusp of the implementation of a more liberal regime of dissection regulations in summer 1933, for example, Roger S. Greene characterized the Beiping police as "conservative, over-cautious,

---

[61] Zhang and Xian, *Minguo yiyao weisheng fagui xuanbian, 1912–1948*, 1–3.
[62] T.D. Sloan to Dr. Houghton, November 14, 1922, CMB, Inc., Box 68, Folder 482, RAC.
[63] Report of Inner Left First District, September 7, 1919, BMA J181-18-10303, 3–7.

and often actually prejudiced, so that even after the permission of the relatives [for an autopsy] has been obtained, the police have often argued with the family in such a way as to lead them to withdraw their consent."[64] While proponents of Western scientific medicine viewed the implementation of dissection as essential for the development of China's modern medical profession, when viewed from the perspective of the police and judicial officials, the benefits, meanings, and implications of dissecting cadavers were undoubtedly more complicated questions. In Beijing, medical schools' claims over the dead body represented the addition of another layer of bureaucratic procedure to the already complex procedures of urban death investigation and the issuing of burial permits. As the staff of schools such as National Medical College and PUMC carried out anatomical dissection, existing practices of death-reporting and investigation had to change in order to accommodate these new claims. In cases such as these, autopsy was not viewed by city authorities as an essential technique in the investigation of deaths or in the surveillance of urban mortality, but, rather, as an activity requiring official supervision, and one that could, if unregulated, challenge the state's existing management of the dead or create societal disturbances stemming from conflicts over the treatment of a body.

Aside from hospital autopsies, bodies were also dissected by medical students during their training in basic anatomy, a part of their education that was meant to develop general knowledge of the body as a foundation for professional knowledge and clinical skills. Procuring the unclaimed bodies of prisoners and the economically marginalized for dissection was a commonly accepted practice in the laws that governed dissection in many European countries, and it formed the basis for the strategy outlined by Tang Erhe and subsequently adopted by National Medical College and other Chinese medical schools.[65] The unclaimed bodies of deceased prisoners, those who had been executed, or those who had died in detoxification and other urban institutions or simply on the streets, thus became a major target of these schools' interest. Gaining access to such bodies required that faculty at PUMC and other schools develop ties with police and judicial authorities, the city officials who were responsible for investigating and burying the bodies of those who died on the margins

---

[64] Roger S. Greene to Alan Gregg, June 21, 1933, CMB, Inc., Box 68, Folder 482, RAC. Also see Roger S. Greene interview with J.R. Cash, August 6, 1927, CMB, Inc., Box 114, Folder 828, RAC.

[65] See, for example, Ruth Richardson's classic study of the 1832 Anatomy Act, which established a framework through which the bodies of those who died in workhouses could be claimed for dissection in England. Richardson, *Death, Dissection, and the Destitute*. For Tang's proposals regarding the sources of cadavers, see Luesink, "Dissecting Modernity," 236–42.

of society. Finding ways to "cultivate" municipal and higher authorities for this cooperation, to use the word of Roger S. Greene, was a common theme in internal discussions among the administration and faculty of PUMC during this period.[66]

The sources of the bodies to which these schools were able to gain access were varied. Over the first decade of the school's operation, the faculty of National Medical College dissected at least thirty-four cadavers, most of which were obtained from the Beijing No. 1 Prison as well as a smaller number of those who voluntarily made their bodies available for autopsy.[67] During the 1920s and 1930s, the school pursued various arrangements with city agencies to provide anatomy and pathology faculty with access to the unclaimed cadavers of those who died unnaturally, in municipal institutions, or which were simply discovered on the streets.[68] In November 1931, for example, the school presented the city's Bureau of Social Affairs with a proposal, originally suggested by Lin Ji himself, that the school provide clinical and hygiene services to workhouses and poor relief facilities in exchange for the unclaimed bodies of those who died in these institutions.[69] By the mid-1930s, the medical school of Beiping University and PUMC had entered into an arrangement in which the schools would alternate in claiming the bodies made available by the police and other agencies.[70] By the late 1930s, this arrangement had regularized both schools' access to a modest supply of bodies.

In Beijing, a basic requirement established by the police was that the bodies of those who died unexpectedly or in poor relief or other institutions could not be claimed by medical schools until after the procuracy's

[66] Roger S. Greene interview with J.R. Cash, August 6, 1927, CMB, Inc., Box 114, Folder 828, RAC; Roger S. Greene to Margery K. Eggleston, December 12 1929, CMB, Inc., Box 68, Folder 481, RAC; C.H. Hu to Roger S. Greene, May 20, 1932, CMB, Inc., Box 114, Folder 829, RAC.

[67] Lin, "Liunian jian bingli jiepou zhi tongji di guancha." For early efforts by National Medical College and PUMC to procure cadavers, see Luesink, "Dissecting Modernity," 241–2.

[68] SJRB December 3, 1926, 6; Beiping Medical University to Beiping PSB, November 3, 1928, BMA J181-20-1185, 8–12. Also see lists of bodies claimed by Beiping University Medical School during the mid-late 1930s contained in BMA J29-3-595. Such lists were routinely submitted to city hygiene authorities as part of their legally mandated oversight of medical school dissections.

[69] Minutes of 91st Meeting of the School Affairs Committee, November 13, 1931, BMA J29-3-8, 149–50; National Beiping University Medical School to Beiping Bureau of Social Affairs, November 27, 1931, BMA J29-3-589, 2–8.

[70] For the original proposal, see PUMC and National Beiping University Medical School to Beiping Municipal Government, December 27, 1934, BMA J29-3-589, 89–94. Lists of cadavers claimed for dissection by Beiping University Medical School during 1936–7 very clearly show this arrangement in operation. See, for example, BMA J29-3-595, 83–7, 104–8.

forensic examination, a reflection of judicial authorities' strong jurisdiction over the bodies of those who died unexpectedly, under suspicious circumstances, or in municipal institutions.[71] We can get a sense of the ways in which police and judicial networks of control and supervision over the dead intersected with the new claims of medical schools from the case of Liang Zhenwen and Li Maolin, detainees at the Detoxification Center for Hard Drugs (Liexing dupin jiechusuo) who died within days of each other in late May 1936 and whose bodies were subsequently claimed by the medical school of Beiping University.[72] Opened in September 1934, this high-profile detention and detoxification facility was run by the Bureau of Hygiene in cooperation with officials of the city PSB, from whom users of opium, heroin, and other illegal narcotics were transferred and to whom they were released again following treatment, at times for an additional period of work in workhouses or other facilities.[73] This facility was one of hundreds of local detoxification centers that were established around the country during the Nanjing decade under the impetus of nationally directed campaigns to suppress the use of illegal opiates.[74] These policies led to mass convictions of drug users that swelled the prison population and, by the mid-1930s, to the summary execution of drug offenders, including those who were caught using drugs after a failed earlier treatment.

On January 17, 1936, the Detoxification Center transferred Liang Zhenwen, a 33 *sui* male who was originally from Shandong, to the city-run Beiping Municipal Hospital, located in the Outer Fifth district, for medical treatment. On May 26, Li Maolin, a 43 *sui* male from Daxing county, was sent to the Bureau of Hygiene's Hospital for Infectious

---

[71] This requirement was formalized in municipal regulations on the claiming of bodies for dissection that went into effect in September 1933, titled "Methods of the Beiping PSB for handling the bodies of those who die unnatural deaths and are executed." For the regulations and related correspondence between the PSB and medical school of Beiping University, see BMA J29-3-589, 39–42.

[72] After claiming Liang's body, the school actually gave it to PUMC for dissection there, as required by the arrangement that these institutions had established for alternating in claiming bodies. See list of cadavers claimed by Beiping University medical school for dissection from the first half of 1936, BMA J29-3-595, 85.

[73] SJRB September 21, 1934, 8; SJRB October 15, 1934, 8; SJRB March 7, 1935, 8; SJRB May 9, 1935, 8.

[74] As Frederic Wakeman has shown, these campaigns occurred within the context of complex moves made at the national and local levels to establish state control over the provision of opiates, including to those who were undergoing treatment for addiction, in order to secure this significant source of revenue. Wakeman, *Policing Shanghai 1927–1937*, 260–75. For the broad effects of Nationalist anti-drug campaigns during this period, see Frank Dikötter, Lars Laamann, and Zhou Xun, *Narcotic Culture: A History of Drugs in China* (Chicago: The University of Chicago Press, 2004), 126–35, 142–5.

Diseases in the Inner Third district after he contracted an unspecified infectious disease. In late May, Liang and Li died within days of each other and the hospitals immediately notified the Detoxification Center of the deaths of the detainees. Given that neither Liang nor Li had relatives to claim the bodies, the Detoxification Center soon offered the bodies of both to the medical school of Beiping University for dissection. By spring of 1936, this facility had joined the network of municipal institutions that were making unclaimed corpses available to the medical school.[75] Many of these bodies were provided by police of the Outer Fifth district, who furnished the medical school with the unclaimed corpses of those who were executed, those who died at the Vagrants' Clinic, a facility discussed in Chapter 1, and some bodies of those who simply died in public places.

At this point, the corpses of Liang and Li were already enmeshed in numerous claims. Most immediately, the Detoxification Center held ultimate responsibility over them, given that they were its detainees. Their bodies were physically located in hospitals in different parts of the city, each of which was accountable to a different police district. Finally, the city procuracy maintained a strong claim over what happened to the bodies given the long-standing requirement that judicial officials inspect the bodies of those who died in detention and other municipal facilities. The Detoxification Center reported Li Maolin's death to officials of the city procuracy, who soon arrived at the Hospital for Infectious Diseases to examine the body. Inexplicably, the Center did not report Liang's death in the Beiping Municipal Hospital – the result, perhaps, of a bureaucratic oversight in the case of a detainee whose transfer to the hospital had occurred months prior. Nonetheless, the medical school of Beiping University soon moved to collect both bodies, despite the fact that, unbeknownst to them, Liang's had been improperly released without a forensic examination.

This bureaucratic error only became known following the discovery of another. The medical school contacted police authorities in order to expedite the dissection of Li's remains, an act that would have contravened a PSB requirement that unclaimed bodies be stored for a month-long waiting period prior to dissection in order to make it possible for relatives to come forward to claim the body.[76] This request was the first

---

[75] Lists of bodies claimed by the medical school in the first half of 1936 indicate that the Center provided the school with three bodies in March of that year. BMA J29-3-595, 73.

[76] See "Methods of the Beiping PSB for handling the bodies of those who die unnatural deaths and are executed," which mandated this waiting period for cases in which it was unknown whether the deceased had relatives who might come forward to claim the body. BMA J29-3-589, 39. It is apparent from the lists of bodies claimed by the medical school that there were lags of several weeks to a month (and more) between the claiming of a body and the dissection procedure itself. There were at least two cases in the first half of

that police authorities had heard about the transfer of Li's body given that, it was now clear, the Detoxification Center had failed to report the death to either the Inner Third district, in which Li had died, or the Outer Third, the district in which the Detoxification Center was located.[77] In investigating the sources of these bodies, police subsequently discovered that Liang's body had never been inspected by judicial officials to begin with – a more serious oversight.

The Beiping PSB headquarters soon contacted the Hospital for Infectious Diseases and the Municipal Hospital and ordered them to notify district police directly when faced with a medical school claiming bodies for dissection in the future.[78] In its reply, the Municipal Hospital noted that in the case of Liang's death, it had followed its usual policy when a patient died: notifying the institution from which the patient had been sent or directly contacting the procuracy to request the forensic examination in the case of the hospital's own patients.[79] In this case, the hospital had followed this policy, contacting the Detoxification Center when Liang died. In its own reply to the PSB, the Hospital for Infectious Diseases included a copy of its internal regulations governing the handling of patients' deaths, which indicated that when a patient died, the hospital's basic procedure was to contact hygiene authorities and the police to have them issue a burial permit.[80] Yet, in this case, the Hospital for Infectious Diseases too had contacted the Detoxification Center rather than notifying police authorities directly. The Outer Fifth district police had been remiss as well in authorizing the transfer of Liang's body from the Municipal Hospital to the medical school without inquiring into whether or not judicial authorities had examined the body first. This oversight also warranted a reprimand from police headquarters in the form of an order that proper procedures be followed in the future.[81]

1936 in which relatives came forward to claim a body that was being stored by the medical school pending the end of this waiting period. BMA J29-3-595, 83–4. In the case of Li Maolin, it appears as though the medical school pathology faculty took interest in autopsying Li's body but was unwilling to delay the procedure, possibly due to concern about the contagious nature of the corpse. In the end, they retained a specimen of Li's skin for further study, photographed the body, and immediately sent it for burial. BMA J29-3-595, 86–7.

[77] Report of Outer Third District, June 2, 1936, BMA J181-26698, 28–31; Report of Outer Third District, June 8, 1936, BMA J181-26698, 32–5.

[78] Beiping PSB to Hospital for Infectious Diseases and Beiping Municipal Hospital, June 5, 1936, BMA J181-20-26698, 9–12.

[79] Beiping Municipal Hospital to Beiping PSB, June 6, 1936, BMA J181-20-26698, 17–20.

[80] Hospital for Infectious Diseases to Beiping PSB, June 10, 1936, BMA J181-20-26698, 21–7.

[81] Beiping PSB to Outer Fifth District, June 10, 1936, BMA J181-20-26698, 2–5.

Conclusion

This bureaucratic tangle reveals several things about the relationships between police, legal authorities, and physicians and, by implication, about the status of physicians' claims over the dead body. For one thing, as important as anatomical dissection was to members of China's medical profession, such claims of scientific authority did not simply or easily translate into claims of legal authority. The anatomical investigations of medical school faculty occurred after city authorities examined the body, a crucial step in the surveillance of urban mortality that was carried out by police and the procuracy in Beijing. In this sense, anatomical dissection was a redundant procedure, the benefits of which were not obvious to the state. It did not directly contribute to the investigative goals of police, procurators, or hygiene authorities. Moreover, as it was implemented in practice, the dissection of cadavers was construed as requiring regulation and supervision. Physicians' claims over bodies for anatomical dissection held the potential to subvert existing mechanisms of police surveillance while making the dead body subject to new interests beyond the family of the deceased, a situation that the police in Beijing perceptively viewed as carrying risk. These competing interests surrounding the dead body were negotiated through the framework of the national and local laws on dissection, a complex ground of inter-professional negotiation in which physicians' claims over the dead body were far from absolute.

Another implication of the cases examined in this chapter is that the implementation of dissection in Beijing did not in any simple sense challenge the authority of the *Washing Away of Wrongs* or the judiciary's forensic claims over the dead. It is not insignificant, for example, that it was the city procuracy that was tasked with investigating PUMC's allegedly illegal autopsy of Song Minghui. In conducting its own forensic examination of a body that had been autopsied in PUMC, this office asserted legal oversight over an institution that maintained great authority within the worlds of Western scientific medicine and medical education.[82] This incident reveals the strong claims that procurators maintained over the dead body well into the 1930s even alongside the institutionalization of Western scientific medicine and its associated programs of anatomical inquiry. Despite forensic reformers' criticisms of the anatomical deficiencies of the *Washing Away of Wrongs*, it was this purportedly unscientific knowledge that police and judicial officials continued to rely on when implementing the state's varied claims over the dead.

---

[82] SJRB August 24, 1930, 7.

This is not to deny the tremendous intellectual and cultural authority that anatomical knowledge attained in early twentieth-century China. Indeed, from the moment that Chinese officials began to observe the medico-legal practices of Japan and other countries during the New Policies reforms, they sought ways of integrating this new kind of body knowledge into the judiciary's existing forensic practices. By the early 1940s, even the forensic personnel of the Beijing procuracy were being instructed in a new anatomical vision of the body as a way of supplementing the forensic knowledge contained in the *Washing Away of Wrongs*.

Ultimately, then, the story told in this chapter is not about official and societal resistance to forensic autopsy – a narrative that is perhaps too easy to accept, given the long history of discourse on Chinese aversion to the opening of the body. Moreover, this chapter has not claimed that officials' reliance on external examination of the body to decide cause of death reflected an inherently conservative resistance toward the adoption of new forensic techniques based on anatomical dissection.[83] Such narratives of resistance to dissection, which have been used to explain the difficulties faced by early twentieth-century proponents of medico-legal reform, fail to take into account the utility that officials derived from their use of systematized practices for examining the surface of the corpse. They also ignore the possibility that such techniques were in fact adequate for determining the manner of death – the question that was, in many cases, the most important one for judicial officials.[84] A more interesting story thus emerges about the ways in which an older set of forensic techniques based on the *Washing Away of Wrongs* informed the decisions of judicial officials about when and how to adopt Western scientific medicine in their cases – one example of the ways in which existing preferences drove the active, critical appropriation of modern science and technology in nineteenth- and early twentieth-century

---

[83] For this assumption, see Jia, *Zhongguo gudai fayixue shi*, 171–2; Jia, "Xinhai geming yihou de zhongguo fayixue," 205–6.

[84] In this connection, see the study of Paul R. Vanatta and Charles S. Petty, who analyzed 185 forensic cases in which a U.S. medical examiner's office conducted an external examination of a body that was later autopsied by hospital pathologists. The study found that an external examination was overwhelmingly accurate when determining the manner of death. Expectedly, this procedure was less capable of accurately diagnosing the medical cause of death in cases involving death from natural causes. To quote the authors, "In a forensic autopsy, the external examination provides much useful information in cases of traumatic death, particularly when supplemented by scene investigation and other evidence. Frequently, internal examination and organ dissection yield little additional information necessary for the completion of a death certificate." Paul R. Vanatta and Charles S. Petty, "Limitations of the Forensic External Examination in Determining the Cause and Manner of Death," *Human Pathology* 18, no. 2 (1987): 172.

China.[85] In this context, as Chen Kuitang suggested in his petition, whether or not it was worth it to physically damage a dead body in pursuit of a new version of forensic modernity – one that was not necessarily superior in practice – was undoubtedly a complicated question.

[85] Reed, *Gutenberg in Shanghai*; Meng Yue, *Shanghai and the Edges of Empires* (Minneapolis: University of Minnesota Press, 2006), 13–30.

Throughout mid–late 1929, procurators in Beiping were in contact with the Hebei High Court regarding a case of suspected poisoning in Hengshui, a county located about 150 miles south of the city.[1] The Hengshui county magistrate had been in contact with Hebei provincial authorities about the case of a deceased individual named Zhang Wenhuan. Zhang's widow, Mrs. Zhang née Yuan, claimed that her husband had been poisoned by a Zhang Wenbo, who had allegedly laced his rice with golden pills (*jindan*), a drug containing morphine or other opiates that was commonly found in parts of north China.[2] Zhang's body had been buried for over a year and authorities in Hengshui wanted to arrange a re-examination of the remains. As was standard practice in these kinds of cases, officials of the High Court now contacted the Beiping procuracy to inquire about the possibility of sending one of the city's inspection clerks to carry out a skeletal examination. In making this request, the Hebei High Court also asked the procuracy to serve as a kind of go-between for authorities in Hengshui and physicians in Beiping who might be able to assist in the case.

The Beiping procuracy subsequently made inquiries with unnamed "medical school and prison physicians," all of whom claimed that it would be fruitless to travel to Hengshui to examine bodily remains in a case that was this old.[3] While in the end this office offered to assist by sending an inspection clerk to examine the remains, it could not help with selecting a suitable physician. In late October the Hebei High Court again contacted authorities in Beiping, this time requesting that they consult with "well-known hospitals like PUMC" about the case in order to

[1] Hebei High Court procurator's office to Beiping Local Court procurator's office, April 25, 1929, BMA J174-2-279, 17–22.
[2] For more on the market for morphine pills in Shanxi during this period, see Henrietta Harrison, "Narcotics, Nationalism and Class in China: The Transition from Opium to Morphine and Heroin in Early Twentieth-Century Shanxi," *East Asian History* no. 32/33 (2006/2007): 166–71.
[3] Beiping Local Court procurator's office to Hengshui County Government, May 3, 1929, BMA J174-2-279, 23–5.

determine if there was a testing protocol that would work in cases involving poisoning by golden pills.[4] The procuracy subsequently contacted seven private and government-run hospitals and medical schools in Beiping with the request.[5]

The replies that were received indicate some of the challenges that judicial officials faced when seeking forensic examination services from medical schools and hospitals during this period. Several of the institutions noted that they lacked specialists in chemical testing and legal medicine, and even the Outer City Government Hospital, an official municipal institution, wrote that forensic testing exceeded the scope of its medical activities.[6] Personnel from the medical school of Beiping University, an institution which by that time had developed a reputation among city authorities for its forensic testing services, demurred on the possibility of dispatching an examiner to Hengshui on the basis that such a trip would cause the school's faculty, who handled forensic testing work, to neglect their classes.[7] In the end, the most substantial reply was given by personnel at the hospital of PUMC, who offered assistance with any queries regarding the principles of toxicological testing, an area of forensic practice for which they would offer actual assistance to city authorities in other cases.[8] They noted, for example, that it might still be possible to test for the poison if it were a metal such as arsenic or mercury, but not if it were an organic substance such as opium.[9] The procuracy duly reported the replies of the medical schools and hospitals to the Hebei High Court. While there is no further correspondence in the archive, one imagines that Hengshui officials' attempts to find a suitable forensic expert ultimately failed.

Within a few years of this exchange, a major change would occur in the accessibility of forensic testing services of the kind that the officials in Hengshui had sought out. By the mid-1930s, judicial officials in counties located in Hebei and throughout China gained unprecedented access to experts in legal medicine, who handled a large number of cases such as

---

[4] Hebei High Court procurator's office to Beiping Local Court procurator's office, October 30, 1929, BMA J174-2-279, 26–30.

[5] Letter of Beiping Local Court procurator's office, November 8, 1929, BMA J174-2-279, 31–4.

[6] For correspondence between these hospitals and the Beiping procuracy, see BMA J174-2-279, 35–44, 48–50.

[7] Beiping University Medical School to Beiping Local Court procurator's office, November 15, 1929, BMA 174-2-279, 51–8.

[8] For more on the PUMC pathology faculty's forensic assistance to city authorities, including a brief mention of their involvement in a case of arsenic poisoning from the early 1930s, see C.H. Hu to Roger S. Greene, May 20, 1932, CMB, Inc., Box 114, Folder 829, RAC.

[9] PUMC hospital to Beiping Local Court procurator's office, November 18, 1929, BMA J174-2-279, 65–70.

this one through an arrangement that relied on the postal system to facilitate the transportation of physical evidence to laboratories in Shanghai and Beiping. This expansion in the availability of forensic testing services was the result of the development of legal medicine as an arm of the state and as a professional discipline, a process of profession-building that occurred under the institutional and financial support of the Nationalist party-state as part of its efforts to promote judicial reform.

Lin Ji, the graduate of National Medical College in Beijing whom we encountered in Chapter 4, was deeply involved in these efforts, serving first as director of the Research Institute of Legal Medicine (Fayi yanjiu-suo), a central government forensic institute in Shanghai, and subsequently in the department of legal medicine of the medical school of Beiping University, a regional center for forensic testing in north China. This chapter examines the "disciplinary program," to use a concept from Timothy Lenoir, that Lin Ji attempted to push forward during the Nanjing decade.[10] The focus will be more on Lin Ji's institutional goals – creating a lasting institutional foundation for the training and employment of experts in the discipline and asserting authority vis-à-vis the judiciary – than on the development of particular areas of research among Chinese legal medicine experts. In fact, one might say that during the 1920s and 1930s legal medicine as a Chinese discipline was more oriented toward establishing an institutional niche for itself than in furthering particular research programs, a reflection of the fact that the field was small and lacked a strong institutional position in Chinese academic institutions and in relation to the judiciary. Transforming the division of labor in forensic examinations was a project that preoccupied Lin Ji.

The development of legal medicine in China during this period involved a process through which proponents of this primarily urban-based scientific discipline attempted to reorganize the institutional and geographic centers and peripheries of Chinese forensics. Beiping itself became an important center of medico-legal testing for north China as officials in Hebei, Henan, Shandong, and Shanxi took advantage of the Beiping University legal medicine department's services. There are parallels, in one sense, with the role that the Beijing procuracy's forensic examiners already played as experts for officials in the region: the city was, yet again, construed as a "center" of forensic expertise to which officials in surrounding counties and provinces might appeal, the assumption being that a greater concentration of expertise existed there than elsewhere.

[10] Lenoir, *Instituting Science*, 53–74.

At the same time, this new arrangement also represented significant differences from the existing organization and conception of the geographic distribution of expert knowledge in forensics. One finds, for example, a new notion of the spatial unevenness of forensic modernity: the idea, as Lin Ji claimed, that the "interior" (*neidi*) of China was a bastion of unscientific forensics.[11] This notion resonated with contemporary claims about the backwardness of medical capabilities in the interior more generally, and suggests the close relationship between histories of modern science, the professions, and urbanization.[12] Lin Ji also viewed the facilities in Shanghai and Beiping as the foundations of a future national infrastructure, rooted in cities with the educational and scientific resources to support medico-legal institutes, which would reorganize China's forensic examination practices. While this ambitious plan did not come to fruition during this period, the Nanjing decade did see a shift through which professional scientists began to exercise unprecedented influence over the forensic work that was carried out in counties throughout China.

The expanding influence of legal medicine could look quite different from these localities, where the *Washing Away of Wrongs* continued to be used in the examination of evidence. For officials who investigated homicide, laboratory testing was easily integrated into an already pluralistic forensic repertoire, not displacing the *Washing Away of Wrongs* but supplementing it as a way of addressing forensic uncertainties generated within this older tradition of forensic knowledge. There was thus a complex negotiation of different forms of institutionalized knowledge that in turn revealed different visions of how forensics should be organized vis-à-vis the state. There was a tension between the imperative, clearly perceived by officials, to continue using the old system of forensic knowledge that had been endorsed by the state and even in the 1930s represented a widely followed standard in Chinese practices of homicide investigation, and an equally compelling drive to adopt the new standards of medicine-based forensics to which China was now held accountable as a modern state and in which it was found to be lacking. In these ways, the story of Nanjing-decade legal medicine is not simply about the successes of a new professional discipline, rooted in urban centers of social and intellectual change, but about an expanded field of encounters between

---

[11] Lin Ji, "Shiyong fayixue" [Practical legal medicine], *Fayi yuekan*, no. 7 (1934): 27; Lin Ji, "Fayixue shilüe" [A brief history of legal medicine], *Beiping yikan* 4, no. 8 (1936): 24; Table of appraisal cases, *Xin yiyao zazhi* 4, no. 5 (1936): 1–3.

[12] Tao Shanmin, "Yiyuan shiyanshi zhi xingzhi gongzuo ji guanlifa" [On the nature, work, and methods of administering hospital laboratories], *Zhonghua yixue zazhi* 19, no. 5 (1933): 714.

distinct traditions of forensic knowledge, each with their own unique claims of effectiveness and authority and each with their own distribution in the administrative geography of the state.

Planning Medico-legal Modernity at the Start of the Nanjing Decade

When we last encountered Lin Ji in Chapter 4, he was pursuing medical studies at Würzburg University following his graduation from National Medical College in Beijing and a short period of employment as an assistant in pathology. When Lin Ji left Germany to return to China in 1928, the country was in the midst of a deep political transformation. The previous year had seen the spectacular military successes of the Northern Expedition, the political bifurcation of the Nationalist revolution in Wuhan and Nanjing, and Chiang Kai-shek's violent repudiation of leftist elements within this erstwhile political coalition. It was not until June 1928 that militarist Zhang Zuolin ceded control over Beijing, a move that brought the former capital under the political influence of the Nationalists and ushered in a decade of decline in the city's status and economic prosperity.[13] Autumn 1928 saw a series of behind-the-scenes political negotiations among Nationalist leaders and their recently acquired militarist allies that, as Ka-che Yip has described, led to the formation of a Ministry of Health under the initial direction of Xue Dubi, a supporter of Feng Yuxiang.[14] Liu Ruiheng (J. Heng Liu), a Harvard-trained surgeon who had just been appointed vice-director of PUMC under the mounting pressure felt by the administration and benefactors of the institution to place Chinese in leadership positions in the school, was tapped to run the Ministry (which was soon reorganized as a subordinate office of the Ministry of Interior), and did so with the assistance of a number of other individuals associated with PUMC.[15]

In late 1928, Lin Ji began a short-term period of appointment under this new central government agency. According to his biographer Huang Ruiting, Lin Ji assisted the Ministry in the formulation of regulations on drugs and pharmaceuticals as well as other medically related legislation, one of a number of early activities that this agency undertook to lay the

---

[13] Dong, *Republican Beijing*, 78–82.

[14] Yip, *Health and National Reconstruction in Nationalist China*, 45–6. For the broader discussions about the nature and extent of the state's involvement in healthcare that were unfolding during the 1920s and that formed an important context for the development of Nationalist health institutions, see Lei, *Neither Donkey nor Horse*, 59–68.

[15] For more on Liu and the broader context of his appointment, see Bullock, *An American Transplant*, 58–60.

groundwork for a renewed and expanded state involvement in healthcare.[16] It is likely that this posting reflected the growing public profile that Lin Ji was attaining during this period. Almost two months before Zhang Zuolin's retreat from Beijing, Lin Ji had published a detailed survey of Beijing's health conditions in the newspaper *Morning Post*, a piece that was intended to catch the attention of a municipal committee that had been formed in March 1928 with the goal of planning a more unified and effective administration of public hygiene in the city.[17] The committee was headed by Shen Ruilin (1875–?), a high-level foreign affairs official of successive Beijing governments during the 1920s whom Zhang Zuolin had invested with authority over the administration of the city. The members of the committee included officials drawn from city and central government agencies, as well as major figures in the field of medical education such as Liu Ruiheng and Yan Fuqing (F.C. Yen, 1882–1970), who was serving at the time in a brief appointment as vice-director of PUMC.[18] Over the next several years, Lin Ji's career would intersect with the professional pathways of both men under the new possibilities created by the Nationalists' early state-building efforts.

During his time assisting central health authorities in Nanjing, Lin Ji also taught at the medical school of Central University, an institution of higher learning established by the Nationalist government in 1927 following reorganization of the former Southeastern University.[19] It was while serving in this position that Lin Ji received a charge from Yan Fuqing, who was dean of Central University's medical school at the time, to formulate a plan for using the school's existing facilities to train experts in legal medicine.[20] The impetus for this proposal had originated with the Central

---

[16] Huang, *Fayi qingtian*, 26–7. For evidence that Lin participated in policy meetings that were attended by Liu Ruiheng and others who would continue to play important roles in planning and building a healthcare infrastructure throughout the 1930s, see minutes of Ministry Affairs Meetings from late 1928, which list Lin Ji as an attendee. *Weisheng gongbao*, no. 2 (1929): *zalu* 5.

[17] The piece was published serially in *Morning Post*: CB April 9, 1928, 7; April 10, 1928, 7; April 11, 1928, 7; April 15, 1928, 7; April 16, 1928, 7; April 18, 1928, 7; April 20, 1928, 7. For the formation of the committee, see SJRB March 20, 1928, 7; SJRB March 29, 1928, 7.

[18] Bullock, *An American Transplant*, 59.

[19] For the politics surrounding the establishment of Central University, see Wen-hsin Yeh, *The Alienated Academy: Culture and Politics in Republican China, 1919–1937* (Cambridge: Council on East Asian Studies, Harvard University, 1990), 120–1. For more on the university's medical school, see Yip, *Health and National Reconstruction in Nationalist China*, 142–3.

[20] Lin Ji, "Niyi chuangli Zhongyang daxue yixueyuan fayi xueke jiaoshi yijianshu" [An opinion regarding the proposed establishment of a legal medicine institute in the medical school of Central University], *Zhonghua yixue zazhi* 14, no. 6 (1928): 205–16.

Political Council (Zhongyang zhengzhi huiyi), which had endorsed an earlier proposal of authorities in Jiangsu. The task was subsequently given to Central University by the University Council (Daxue yuan), a short-lived central government office for administering national education policy that was run by the former chancellor of Peking University and pre-eminent educator Cai Yuanpei before its precipitous disbandment in fall 1928.[21] This proposal represented one of several attempts made by officials in the lower Yangtze during the first few years of Nationalist rule to establish provincially or nationally funded legal medicine training institutions. In 1930–1, for example, a short-lived training program was established by the Tongde Medical School in collaboration with Jiangsu High Court for purposes of providing medical school graduates with specialized training in legal medicine.[22]

At the core of Lin Ji's proposal was the establishment of a department of legal medicine within the school of medicine of Central University. In proposing such an institution, Lin Ji was following a model that, by the late 1920s, could be observed in many European countries, Japan, and elsewhere – namely, the establishment of "institutes" of legal medicine, independent laboratories or shared university facilities that housed the equipment and facilities necessary to support this discipline of scientific medicine.[23] Variable in scale, capital investment, and government support, these institutions commonly supported a range of activities that included the training of medical students, police, and judicial authorities; the examination of evidence sent from police and legal authorities, including the storage, identification, and autopsy of bodies; and academic research. The institute-centered model of forensic investigation and research was not universally adopted. Norman Ambage and Michael

---

Also see discussion of the proposal in Huang Ruiting, *Zhongguo jinxiandai fayixue fazhan shi* [A history of the development of legal medicine in modern China] (Fuzhou: Fujian jiaoyu chubanshe, 1997), 48–55.

[21] For the history of the University Council, see Allen B. Linden, "Politics and Education in Nationalist China: The Case of the University Council, 1927–1928," *The Journal of Asian Studies* 27, no. 4 (1968): 763–76.

[22] See brief mention of this program in Yang, "Fayixue shilüe bu," 478. For regulations governing the program, see the document appended to a letter from Hebei High Court to National Beiping University Medical School (June 27, 1930) transmitting Jiangsu authorities' solicitation for enrollees. BMA J29-3-587, 7–8.

[23] For an overview of the history, organization, and activities of a number of medico-legal facilities, see The Rockefeller Foundation, Division of Medical Education, *Methods and Problems of Medical Education (Ninth Series)*. For the process through which nineteenth-century medical disciplines were reorganized as specialized fields of experimental science requiring autonomous laboratory facilities, see Richard L. Kremer, "Building institutes for physiology in Prussia, 1836–1846: Contexts, interests and rhetoric," in *The Laboratory Revolution in Medicine*, ed. Andrew Cunningham and Perry Williams (Cambridge: Cambridge University Press, 1992).

Clark have shown, for example, that this model was less suited to the administration of justice and policing in early twentieth-century England than were police science laboratories that cost less and appeared to be less rooted in a narrowly medicine-based forensics.[24]

The institute that Lin Ji envisioned for Central University would support three related sets of activities: an advanced training program for medical school graduates, a program for training a larger number of forensic assistants, and a department that would provide forensic services to judicial authorities requiring assistance in their cases.[25] Initially, Lin Ji proposed a staff of about a dozen, including professors, lecturers, assistants, technicians, and a secretary. These personnel would be responsible for providing instruction to classes of 40–50 researchers and 50–60 assistants, all while handling forensic examination work in cases sent from courts. The instruction provided to both groups would encompass the broad range of medical and scientific disciplines on which legal medicine was based, ranging from foundational classes in anatomy and pathology to toxicology, forensic psychiatry, and other specialized technical subjects. The researchers would also take courses in social medicine, administrative medicine (*xingzheng yixue*), life insurance medicine, disaster medicine, and medical legislation. In all, creating these institutions would cost approximately 10,000 yuan in start-up costs, with recurring monthly costs of about 2,000 yuan for the salaries of instructional staff and testing materials.

More than simply a plan for establishing these particular institutions, this proposal was in fact a blueprint for building a new profession and embedding its members in state institutions. Lin Ji envisioned Central University's legal medicine department as training a cadre of specialized personnel who would be distributed in time to courts and police agencies across the country. These experts would handle the examination of forensic evidence in legal cases while attending to broader matters of public hygiene and medical administration, a vision that was undoubtedly inspired by Germany's system of local medical officers.[26] Once established in local jurisdictions, experts trained at the proposed facility would "carry out new-style forensic examinations while establishing local medical laws and ordinances and investigating the suffering of society."[27]

---

[24] Norman Ambage and Michael Clark, "Unbuilt Bloomsbury: Medico-Legal Institutes and Forensic Science Laboratories in England between the Wars," in *Legal Medicine in History*, ed. Michael Clark and Catherine Crawford (Cambridge: Cambridge University Press, 1994).

[25] Lin, "Niyi chuangli Zhongyang daxue yixueyuan fayi xueke jiaoshi yijianshu."

[26] For a description of the role, training, and testing of medical officers in German states during this period, see Gottfried Frey, *Public Health Services in Germany* (n.p.: League of Nations, Health Organisation, preface 1924), 25–30.

[27] Lin, "Niyi chuangli Zhongyang daxue yixueyuan fayi xueke jiaoshi yijianshu," 207.

The latter might entail forays into the realms of public hygiene, industrial hygiene, and other fields that intersected with medico-legal research and practice. Lin Ji's own research into the medical pathologies and substance abuse afflicting thirty-six of Beijing's rickshaw pullers, for example, indicated how medical research might be used to promote the well-being of society. In this instance, Lin Ji's detailed physical examination of the pullers had formed the basis for his proposal that the young and old be prohibited from engaging in this occupation and that all rickshaw pullers be subjected to regular health inspections.[28]

Those who advocated physicians' involvement in legal cases had been arguing vocally for the medical professionalization of forensic investigation since the 1920s, as we have seen. Lin Ji's proposal simultaneously promoted an expansion of the role of physicians in the law while restricting which medical professionals could serve as the most authoritative forensic experts. The ideal forensic practitioner would be a medical school graduate who had received two additional years of training. This expert role would be formalized several years later with the creation of a new credential, that of "medico-legal physician" (*fayishi*), the possession of which was meant to distinguish physicians who had received advanced training in legal medicine from those who had not. The particular strategy that Lin Ji advocated for reforming forensics thus relied on the assumption that "ordinary physicians" (*yiban yishi*), in his words, would be ill-equipped to handle specialized areas of forensic practice such as toxicological testing and blood analysis.[29] Lin Ji was thus arguing for the distinctively specialized nature of legal medicine as a medical discipline and the importance of prioritizing advanced post-graduate training as a remedy for the deficiencies of China's forensic capabilities.

This proposal represented an ambitious vision for the formation of a new profession and the creation of new relationships between medicine and law, science and governance. It envisioned the legal medicine institute as the intersection between the professional fields of education, law, and medicine and as the source of a new national infrastructure composed of medico-legal officials and assistants stationed in every county and municipality. Once additional facilities had been established in Shanghai, Beiping, Hankou, Guangzhou, Chongqing, and Fengtian (Mukden), Lin Ji argued, specialized medico-legal services would be available to local authorities throughout China.[30] This hierarchically

---

[28] Lin Ji, "Renlichefu xinzang ji maibo zhi biantai" [Abnormalities in the Heart and Pulse of Rickshaw Pullers], *Zhonghua yixue zazhi* 14, no. 4 (1928): 252–70. For Lin's proposals regarding the regulation of this occupation, see CB April 18, 1928, 7.

[29] *Xin yiyao zazhi* 4, no. 5 (1936): 2.

[30] Lin, "Niyi chuangli Zhongyang daxue yixueyuan fayi xueke jiaoshi yijianshu," 216.

organized national distribution of forensic experts would be rooted in urban centers, locations that had the educational facilities and resources to support specialized training and investigation work in this field. This particular formulation of forensic reform as relying on the training of a new group of examiners and their distribution to local jurisdictions also implied tremendous expenditures and new (albeit unspecified) ways of finding financial resources for these personnel within the tight judicial budgets of county and provincial governments. Lin Ji's vision of forensic modernity thus ignored the possibility that the judiciary's existing forensic personnel could be retrained according to the evidence collection and examination standards of legal medicine, a strategy that would have required at least temporarily accepting the judiciary's existing forensic personnel and practices as a transitional step toward a physician-centered forensic examination system. While Lin Ji's proposal for using Central University as the foundation for a broader series of reforms was never adopted, the basic assumptions underlying this plan informed many of his subsequent efforts during the 1930s.

The Research Institute of Legal Medicine

Soon after making this proposal, Lin Ji returned to the medical school of Beiping University with an appointment as directing professor of the Institute of Legal Medicine, a department that was now independent from pathology. During the period of Lin Ji's involvement with the Ministry of Health and Central University, Xu Songming had handled administrative matters pertaining to legal medicine at the school, an area of instruction and practice in which he had long been involved.[31] Upon his return to the school, Lin Ji embarked on an ambitious campaign to contact central and regional judicial authorities throughout China and notify them of the forensic services offered by the school's medico-legal laboratory, a strategy that he would repeat again later in the 1930s.[32] Lin also formulated a proposal similar to the one made in 1928 for using the medical school of Beiping University as the foundation for new training and investigation facilities and to begin the process of creating a national infrastructure of newly trained forensic experts.[33] The proposal received

---

[31] Minutes of School Affairs Assembly, December 26, 1928, BMA J29-3-1, 38; Minutes of School Affairs Assembly, November 14, 1930, BMA J29-3-3, 189–90.

[32] Correspondence between the medical school and the Judicial Yuan, Hebei High Court, and other judicial organs can be found in BMA J29-3-587, 14–21, 28–39.

[33] The proposed institutions would be funded jointly by Beiping University medical school and central and provincial judicial organs. For the proposal as well as the Ministry's reply, see *Sifa xingzheng bao*, no. 4 (1932), *gongwen* 21–6. Also see "An opinion regarding

a lukewarm response from officials at the Ministry of Judicial Administration, who cast it as largely redundant given that the medical school of Beiping University was ostensibly already capable of providing advanced training in legal medicine, courts in north China had already been ordered to send cases to this institution, and training programs more economical than the one which Lin Ji proposed had already been held by authorities in Jiangsu.[34]

When Lin Ji drafted this proposal for the north China medico-legal institutions, he was aware that the Ministry of Judicial Administration had already begun planning a legal medicine training facility to be built under its own auspices in Zhenru, a town to the northwest of the International Settlement in Shanghai. In fact, within two months of the Ministry's rebuff of Lin Ji's proposal to expand legal medicine training in north China, this agency would turn to him for assistance in completing its new forensic training institution in Shanghai. Initially, the Ministry had given the task of planning this facility to Sun Kuifang, a graduate of Peking University who had pursued doctoral studies in medicine at the University of Paris, where he had also undertaken advanced training in legal medicine.[35] From 1930–1, Sun carried out various activities required for planning this institution, purchasing land in Zhenru and even traveling to Europe to purchase books and instruments.[36] In the end, however, Sun would not see the facility to completion.

There are several possible reasons why central judicial authorities asked Lin Ji to complete the preparations for the Ministry-sponsored institute in lieu of Sun in mid-April 1932. While planning the facility, Sun Kuifang handled forensic investigation work for judicial officials in Shanghai. It was during this period that Sun was forced to defend his involvement in the case of Xuan Axiang, a woman who had died from accidental suffocation yet whose death was initially suspected to have been a homicide. Judicial authorities had made this erroneous assumption in part on the basis of evidence provided by Sun himself. Beyond this case, Sun seems to have had other problems with his public image. For

preparations to establish medico-legal research and forensic inspection organs in Beiping," August 1931, BMA J29-3-71, 4–11.

[34] Ministry of Judicial Administration to Beiping University Medical School, February 25, 1932, BMA J29-3-587, 45–8.

[35] For basic biographical information on Sun Kuifang, see *Fayi yuekan*, no. 16 (1935), 75.

[36] *Sifa xingzheng bu Fayi yanjiusuo choubei jingguo qingxing ji xianzai chuli shiwu ji jianglai jihua gailüe* [An account of the course and circumstances of preparations for the Research Institute of Legal Medicine of the Ministry of Judicial Administration and summary of work presently handled and future plans] (Shanghai: Fayi yanjiusuo, 1932), 1–2; "Fayi yanjiusuo chengli zaiji" [The Research Institute of Legal Medicine will be established shortly], *Zhonghua faxue zazhi* no. 8 (1931): 101–2.

example, while defending his involvement in the Xuan Axiang case, Sun also felt the need to publicly address damaging allegations that he "[had taken] a huge sum of money and went abroad and [preparations for] the Shanghai medico-legal inspection institute subsequently vanished without a trace."[37] Sun's preparations for the Research Institute were also clearly behind schedule by mid–late 1931, a delay that was compounded when Japanese soldiers occupied northern districts of Shanghai, including Zhenru, after an attack on the city in early 1932.[38] It was only after May, when Zhenru was returned to Chinese control, that the land for the facility was reclaimed. By this point, the Ministry had already appointed Lin Ji head of planning and first director.

On August 1, 1932, the Research Institute of Legal Medicine of the Ministry of Judicial Administration formally opened. The facility was located on a piece of land abutting the campus of Jinan University and located along the Nanjing-Shanghai railway line in northwest Shanghai. The scale of this facility far surpassed those of Lin Ji's earlier proposed institutions at Central University and Beiping University Medical School, which were conceived as part of school facilities rather than as stand-alone institutions.[39] The Research Institute, by contrast, occupied a number of buildings within a walled compound. It had laboratories for the testing of poisons and blood, facilities for autopsy and histological examination, rooms for examining witnesses and conducting psychological assessments, and morgue facilities for storing bodies. It also contained a darkroom, a library, a specimen collection room, and the equipment and books to support academic research in the various fields of medico-legal inquiry. The personnel organization of the Research Institute included a director; two section chiefs responsible for overseeing all areas of the facility's investigation, training, and research activities; a general manager responsible for the day-to-day administration of the facility; about thirty technical staff members and assistants; and instructional staff to provide training to the classes of institute researchers, as

---

[37] "Xuan Axiang an jianyan jingguo," 29.

[38] "Fayi yanjiusuo chengli zaiji." According to the brief history of the Research Institute's planning provided in Lin Ji's first year work report, in the end the preparations were "left uncompleted after a long time" despite Sun's initial efforts. Lin, "Sifa xingzheng bu fayi yanjiusuo chengli yizhounian gongzuo baogao," 1. For a detailed account of the tensions leading to the attack, the destruction wrought on Zhabei and the northern parts of Shanghai, and Chinese efforts to regain jurisdiction over these areas from the occupying Japanese, see Wakeman, *Policing Shanghai 1927–1937*, 181–212.

[39] *Sifa xingzheng bu Fayi yanjiusuo choubei jingguo qingxing ji xianzai chuli shiwu ji jianglai jihua gailüe*; Lin, "Sifa xingzheng bu fayi yanjiusuo chengli yizhounian gongzuo baogao," 2–3. As of early 1934, the monthly operating costs of the facility reportedly amounted to 5,500 yuan. Zhuang Wenya, *Quanguo wenhua jiguan yilan* [An overview of the country's cultural organs] (Shanghai: Shijie shuju, 1934), 240.

well as outside experts whom the Research Institute could not afford to employ full-time yet still brought on as "honorary technical specialists" (*mingyu jishu zhuanyuan*).[40]

Soon after the Research Institute was established, Lin Ji set goals for training personnel that were similar to his earlier proposals made in 1928 and 1931. This involved the proposed creation of a tiered cadre of specialist researchers and assistants who would be stationed in provincial and local courts following completion of advanced training at the Research Institute. By May 1933, the Research Institute was seeking medical school graduates to enroll as researchers with the intention of beginning training in August of that year.[41] About a week after the official registration period began, Lin Ji contacted officers of the medical school of Beiping University and requested their assistance in administering admissions testing in Beiping to those for whom it might be difficult to travel to Shanghai.[42] Of the seven who completed enrollment testing in Beiping, six were accepted by the Research Institute to join the class or serve as alternate candidates; in the end, of the seventeen individuals who would complete the 18-month training program, four were graduates of Beiping University Medical School.[43]

Each graduate of the class received a certificate conferring the credential of "medico-legal physician" (*fayishi*), signed by Wang Yongbin (1881–?), an early follower of Sun Yat-sen who had recently become head of the Ministry of Judicial Administration following a number of significant political appointments under the Nationalist government. This was a new professional title, purposefully created to distinguish the graduates of the program from the *fayi* (legal medicine experts) who were already employed by some law courts. Because there was no uniformity in the professional background of those who held the latter title, Lin Ji argued to officials of the Ministry, it was now necessary to establish a new credential to distinguish those who had received the superior training. The credential was also meant to demonstrate that forensic

---

[40] *Sifa xingzheng bu Fayi yanjiusuo choubei jingguo qingxing ji xianzai chuli shiwu ji jianglai jihua gailüe*, 6–7; Zhuang, *Quanguo wenhua jiguan yilan*, 240; "Fayi yanjiusuo juxing erzhou jinian Lin suozhang baogao jingguo," 1101.

[41] See, for example, the solicitation letter that Lin Ji sent to Beiping University Medical School on May 11, 1933. BMA J29-3-587, 68–73. Soon after the Research Institute's founding, Lin Ji had described the creation of such a training program in a list of goals to be accomplished in 1933. *Sifa xingzheng bu Fayi yanjiusuo choubei jingguo qingxing ji xianzai chuli shiwu ji jianglai jihua gailüe*, 13.

[42] For correspondence relating to this arrangement, see BMA J29-3-587, 84–7, 93–6, and 100–8.

[43] A list of the first class of researchers and their biographical information was published in *Fayi yuekan*, no. 10 (1934).

examiners were more than simply physicians, but rather were specialists who had undergone additional post-graduate training:

Because the researchers who graduate from our institute both have the qualifications of physicians and have pursued additional specialized study, adding a "law" character in front of the word "physician" (*yishi*) to create the title "medico-legal physician" (*fayishi*) and making this the credential received upon graduation seem extremely appropriate. According to the terminology used for medical specialists, those physicians who pursue additional specialized studies in surgery, internal medicine, obstetrics, or gynecology are then called "physician of surgery," "physician of internal medicine," or "physician of obstetrics and gynecology." Calling those who specialize in legal medicine "medico-legal physician" also seems appropriate.[44]

The creation of this new credential was meant to suggest that legal medicine specialists possessed expert knowledge that surpassed that of other physicians of Western medicine. It also reflected Lin Ji's interest in developing an elite tier of medico-legal experts as the first, most important step in transforming the forensic practices of local courts. It is important to acknowledge, of course, that this new credential did not, in any way, allow the legal medicine profession to actually restrict who could serve as a forensic examiner in Chinese courts and county judicial offices. One did not have to have undergone advanced training at the Research Institute to serve as an outside expert in legal cases or to be employed by local judicial authorities as a forensic examiner. Moreover, this new credentialing scheme did nothing to address, restrict, or otherwise impact the judiciary's inspection clerks, the forensic practitioners who collected much of the evidence that was used in Chinese courts.

Nonetheless, the graduation of the first class of such specialists in late 1934 represented an unprecedented development: uniformly trained forensic professionals could now be distributed nationally, a significant shift from the largely ad hoc training efforts that had been carried out in local settings since the 1920s. Of the seventeen who completed the program and received this credential, fourteen were sent to posts in provincial high courts where they were to handle, among other work, the re-examination of questioned evidence in cases sent from counties.[45] While, as Lin Ji acknowledged, the number of those graduating was "still

---

[44] In late May 1934 the Ministry of Judicial Administration sought the Examination Yuan's approval for the Research Institute's scheme to establish this special credential for those who graduated from the advanced training program. *Kaoshiyuan gongbao* no. 6 (1934): *gongwen* 22.

[45] For a list of the program graduates and the institutions to which they were dispatched, see *Fayi yuekan*, nos. 12/13 (1935). These included the provincial courts of Guangdong, Shandong (3 individuals), Zhejiang (2 individuals), Hebei, Jiangsu (2 individuals), Henan, Sichuan, Hunan, Guangxi, and Hubei.

not enough for distribution to the courts of the entire country," the program had positioned highly trained experts in regional judicial offices, an initial step in the eventual goal of influencing the forensic practices carried out in all localities.[46] Once distributed in this way, these experts did become involved in local cases even though their impact was undoubtedly limited when compared to the more widespread influence of local courts' existing forensic examiners.[47] According to data collected in a national survey of judicial institutions carried out in 1929, before the establishment of the Research Institute, out of 706 individuals who were working as forensic inspection personnel in courts and local judicial offices, 291 (41 percent) were identified as being "*wuzuo*" – ostensibly, individuals who had served since the late Qing – while another 268 (38 percent) were new forensic personnel who had been trained informally or in apprenticeships with local judicial personnel.[48] Only 147 (21 percent) had received formal, school-based training in forensic examination, seemingly in one of the ad hoc training classes for inspection clerks that had been held periodically under the late Qing and early Republic. In this context, while dispatching a dozen or so medico-legal experts to provincial courts did represent the first steps of a particular kind of forensic reform, it did not change the ways in which most local judicial offices organized their forensic capabilities or the kinds of body examinations that were carried out in their cases.

Science through the Mail

In early 1935, Lin Ji resigned from his position as director of the Research Institute. The public notice published in *Monthly Bulletin of Legal Medicine* claimed that this was because of a duodenal ulcer, a health condition that required time to convalesce without the heavy burdens of directing this central government institute.[49] Given that Lin Ji's return to the medical school of Beiping University soon found him engaged in the same intensive pace of work, albeit on a smaller scale, it is possible that this was a public pretext necessitated by his resignation under less than

[46] *Fayi yuekan*, nos. 12/13 (1935), *xu*.

[47] For example, for a case of suspected poisoning in which the Research Institute-trained expert Cai Jiahui was called to examine the body after the initial examination by a local inspection clerk, see Beiping University Medical School Institute of Legal Medicine, Written appraisal report, April 17, 1936, BMA J29-3-608, 37–8.

[48] *Quanguo sifa huiyi huibian* [Collected materials of the National Judicial Conference] (n.p.: 1935), *yian, di 4 zu*, 24–6. Also see Ming Zhongqi, "Woguo fayi qiantu de zhanwang" [A view of the future prospects for legal medicine in our country], *Dongfang zazhi* 33, no. 7 (1936): 185.

[49] *Fayi yuekan*, no. 14 (1935): 76.

favorable conditions.[50] At any rate, Lin Ji soon found himself engaged intensively in the other aspect of profession-building with which he had been involved at the Research Institute: handling forensic cases for local judicial officials. This too was an essential activity for demonstrating the value of medical expertise to Chinese legal officials, albeit one that was carried out at a distance through the postal exchange of physical evidence and appraisal reports. In the end, this arrangement would become one of the most important modalities through which medico-legal experts expanded their influence among legal officials during the 1930s.

One of Lin Ji's accomplishments during this period was to regularize county-level officials' access to forensic testing services. In the period between the Research Institute's opening in August 1932 and July 1933, aside from handling cases from authorities in Jiangsu (70 cases) and Zhejiang (4), it also investigated cases from Shandong (5), Hubei (4), Hebei (3), Guangxi (3), Anhui (3), and Sichuan, Jiangxi, and Hunan (1 case each).[51] That in summer 1934, personnel of the Research Institute were examining evidence in a case of suspected poisoning that had originated in eastern Gansu (discussed below) indicates the extent to which the facility had already expanded access to forensic laboratory testing.[52] After Lin Ji returned to Beiping in 1935, he similarly expanded access to testing services for county-level officials in the four provinces of Hebei, Henan, Shandong, and Shanxi, a region of the country in which judicial institutions were underdeveloped in comparison to those of the lower Yangtze and south China.[53] Between April 1935 and late March 1936, the Institute handled 35 cases from these provinces involving suspected blood stains (in 19 of these cases, the stains turned out to be human blood, but in the rest they were not blood stains) and

---

[50] This is suggested by Chen Kangyi, who remained at the Research Institute in early 1935 after completing the training program. Aside from noting that Lin Ji was in fact quite ill during this period, Chen writes: "After Wang Yongbin took the post of head of the Ministry of Judicial Administration, there were many personnel changes. For example, the directors of provincial high courts were transferred around. The Research Institute of Legal Medicine was no exception. Director Lin left the research institute that he personally founded to return to teach at Beiping University Medical School." Chen Kangyi, "Daonian wo jing'ai de laoshi – Lin Ji jiaoshou" [Mourning my revered and beloved teacher – Professor Lin Ji], *Zhongguo fayixue zazhi* 6, no. 4 (1991): 234.

[51] Lin, "Sifa xingzheng bu fayi yanjiusuo chengli yizhounian gongzuo baogao," 6–7.

[52] *Fayi yuekan*, no. 11 (1934), 130–2.

[53] As Glenn Tiffert shows, Hebei province, from which a number of forensic cases were sent to Beiping, was deficient in almost every metric of judicial institution-building, including the numbers of courts and county judicial offices (in which the authority to adjudicate was held by an official who was different from the county magistrate, who still served as prosecutor), and the capacity to accept civil cases and bring them to resolution. This situation had not changed significantly by the late 1940s. Tiffert, "An Irresistible Inheritance."

43 cases involving suspected poisoned food (51 percent contained poison, all of which was arsenic), in addition to a number of other kinds of cases.[54]

This arrangement removed the uncertainty that officials such as those in Hengshui faced when seeking the assistance of medical experts in their cases. Officials did not have to contact unfamiliar hospitals and medical schools about specialized forensic services that they were in all likelihood ill-equipped to provide. It made advanced laboratory testing services accessible to judicial offices located in parts of the country that lacked the resources necessary to support the development of such institutions, and it relieved officials of having to seek funds to employ a new forensic expert in the county or Local Court. Rather, officials simply chose the testing services that matched their concerns in the case at hand. The costs associated with these services varied greatly. Regulations of Lin Ji's forensic laboratory in Beiping specified, for example, the costs for identifying the type and quantity of toxins in food ingredients or on cooking implements (5 yuan), in cooked foods (10–20 yuan), or in medicines (10–50 yuan); investigating signs of poisoning in organs that had been removed (6–25 yuan) and entire bodies (10–30 yuan); or testing skeletal remains for metallic poisons (20–50 yuan). It was the policy of this institution to reduce prices for officials who were investigating criminal cases. In a case sent from Shandong High Court in 1936, for example, the cost of determining whether a sample of food contained poison was reduced from 5 to 3 yuan.[55] The wide use that officials made of these services suggests that the costs were within the range of local judicial budgets and that officials saw the services that they did request as worth the expense.[56]

---

[54] *Xin yiyao zazhi* 4, no. 5 (1936), table of cases.

[55] "Provisional measures for appraisals and examinations," October 1935, BMA J29-3-607, 3–10; Shandong High Court to Beiping University Medical School, May 28, 1936, BMA J29-3-608, 306–9.

[56] We might compare these costs to those associated with the judiciary's skeletal examination practices. In such cases, local officials were responsible for the costs associated with inspection clerks' travel from Beijing and purchase of the supplies necessary to carry out a steaming examination. In one case, involving a skeletal examination in Xinle county, Hebei, the cost of having Yu Yuan and Yu Delin carry out the examination included travel expenses of 15 yuan and daily board and lodging costs of 1.5 yuan per person. Procurator's office of Beiping Local Court to Xinle County Government, September 6, 1929, BMA J174-2-279, 92. There were various ways that the costs involved in such examinations could be reduced. For example, if a local inspection clerk could assist in the examination, the Beijing procuracy might only have to send one assistant along with the expert inspection clerk instead of two, thereby reducing travel costs and other expenses. Expenses could also be cut significantly in cases for which only single bones, not the entire skeleton, were examined. Thus, when investigating

Given these arrangements, one can conclude that this period saw an expansion of the professional authority of experts in legal medicine. At the same time, while the increased use of medico-legal laboratory testing did represent a significant shift in the influence and status of this discipline in China, the result was not an arrangement in which physicians gained an *exclusive* jurisdiction over forensic examinations. This was the case because local officials continued to handle the initial examination of bodies and other evidence. Moreover, it was not an arrangement in which officials had completely abandoned the forensic practices of the *Washing Away of Wrongs* in favor of those associated with the laboratory, a fact that presented challenges for forensic experts who worked in laboratories at a distance from the localities in which the evidence was actually collected. Thus, as interactions between China's first forensic laboratories and local judicial offices multiplied, the result was less the emergence of a new forensic infrastructure based on medical expertise than a pluralistic arrangement in which homicide was investigated at the intersection of law and medicine, with physical evidence examined using different techniques and standards of proof.

The View from Local Courts

In late May 1934, the Research Institute of Legal Medicine received a request for forensic laboratory testing from the High Court of Gansu province.[57] The request pertained to physical evidence in a case that had originated in Jingchuan county, an area of eastern Gansu located about one thousand miles from Shanghai. A local person of Xijin village had accused the village head and another local official of beating his younger brother to death. Because the county magistrate, who also served as prosecutor in the county judicial office, was ill at the time, subordinate officials were sent to conduct the forensic examination. These officials not only found wounds on the body of the deceased, but also discolored skin and livid fingernails that were consistent with the signs of fatal poisoning. In interpreting these as evidence of poisoning, the officials were following guidelines included in the *Washing Away of Wrongs*, which contained detailed descriptions of the skin discolorations, swellings, and discharged blood and excreta that one could expect in cases of fatal poisoning.[58]

officials proposed that the re-examination in the case of Liu Guangju only focus on four ribs, Yu Yuan explained that no more than a third of the necessary distiller's grains, spirits, and vinegar would be needed. Xinle County Government to procurator's office of Beiping Local Court, August 25, 1929, BMA J174-2-279, 87; Report of Yu Yuan, June 11, 1923, BMA J174-1-184, 73.

[57] *Fayi yuekan*, no. 11 (1934), 130–2.   [58] *Lüliguan jiaozheng Xiyuan lu*, 3.17a.

In late imperial China, much as in early modern England, it was generally believed that the fact of fatal poisoning could be confirmed on the basis of easily observed external signs such as these. Indeed, much as Ian Burney has noted for the interpretation of bodily signs of poisoning in the latter context, one might say for the methods of confirming poisoning in the *Washing Away of Wrongs* as well that "the value of surface signs of poisoning available to the trained eye and the casual onlooker alike was a matter widely accepted in learned texts and in everyday practice."[59]

Having observed these signs on the body, the officials now wanted to use another method for confirming poisoning that was described in the *Washing Away of Wrongs* – namely, inserting a silver needle into the dead body and looking for the dark discoloration that would appear on the needle, it was believed, if it came into contact with traces of poison in the body. In the most basic scenario, officials confronted with a case of suspected poisoning were instructed to obtain a silver hairpin and wash it carefully in a solution containing water and pods of the soap-bean tree.[60] They were then to insert and seal the hairpin in the mouth, throat, or rectum of the deceased, the location chosen to afford maximal contact with the poisonous matter still in the body. Upon removal, officials were to inspect the hairpin for a bluish-black (*qinghei*) color that would indicate that it had come into contact with poison. Late imperial commentators expressed confidence in the technique, and cases that were appended to expanded editions of the *Washing Away of Wrongs* demonstrated that in instance after instance the needle would darken when inserted into the body of a person who had been poisoned.[61]

We can explain the use of this forensic test on the basis of the wider belief in late imperial China that silver would change color when exposed to poison, a phenomenon that the critical and curious naturalist Li Shizhen (1518–93) seemingly took for granted, and one that informed the Qing imperial palace's use of dishes made of silver, or that featured a visible silver strip, as a means of detecting poison.[62] Modern critics, including Lin Ji, condemned Chinese officials' use of the test given that any discoloration of a silver object inserted into a corpse could simply be explained as silver sulfide staining or, to put it simply, that the silver tarnished upon contact with sulfur-containing chemical by-products of

---

[59] Burney, *Poison, Detection, and the Victorian Imagination*, 51.
[60] *Lüliguan jiaozheng Xiyuan lu*, 3.17b-3.19a.
[61] *Chongkan buzhu Xiyuan lu jizheng*, 3.44a-45b.
[62] Li Shizhen, *Bencao gangmu: Xin jiaozhu ben* [Systematic Materia Medica, A new critically-annotated edition], edited by Liu Hengru and Liu Shanyong (Beijing: Huaxia chubanshe, 2002), *juan* 8, *yin*, v. 1, 329. The latter practice is mentioned in Aisin-Gioro Pu Yi, *From Emperor to Citizen – The Autobiography of Aisin-Gioro Pu Yi*, translated by W.J.F. Jenner (Peking: Foreign Languages Press, 1979), v. 1, 43.

bodily decomposition.[63] Thus, in their estimation, this could hardly be used as a reliable test for fatal poisoning. In fact, in cases of suspected poisoning, late imperial officials used the test in conjunction with other forensic evidence such as discolored skin or other signs of the damage caused by a toxin in the body.[64] The use of a silver hairpin or needle to confirm fatal poisoning thus constituted one among several sources of evidence that could be used, including other tests described in the *Washing Away of Wrongs* such as feeding rice that had been placed for a time inside the throat of a suspected poisoning victim to a chicken. One might speculate as well that officials used this test to pressure suspects to confess by appealing to the seemingly irrefutable evidence of poisoning that it produced, demonstrated so clearly in a dark discoloration on the hairpin or needle that could not be wiped away. In this sense, one might compare the function of the silver needle test to that of the lie detector, an invention of twentieth-century American policing that is effective at eliciting confessions in part because the person being interrogated believes in the technological certainty of the machine itself.[65]

In the case involving the Xijin villager, this test was not initially carried out. As the investigation began to indicate that the death had been caused by poisoning, the relative of the deceased caused an uproar and persisted in claiming that it was a case of fatal beating, thereby preventing the officials from performing the test. When another county official arrived in the village to carry out the test, the relative still refused to allow the procedure to go forward. In the end, the county magistrate personally traveled to the village, despite his illness, and persuaded the relative to allow the test, which was subsequently carried out.[66] When the needle was removed from the now 10-day old corpse, it had black marks that the relative was allowed to scrub with a solution of water and soap pods. This post-removal scrubbing was crucial given that, as the *Washing Away of Wrongs* explained, a silver hairpin could blacken when exposed to the putrid fumes (*huiqi*) of a corpse. While this kind of discoloration, caused by exposure to putrefaction, could be expected to disappear after being

---

[63] For modern assessments of this test, see discussion of Lin Ji's experiments below, as well as Tang Tenghan, "Xiyuan lu shang zhi huaxue wenti" [Chemistry problems in the *Washing Away of Wrongs*], *Guofeng banyuekan* 3, no. 12 (1933), 21–3. For a positive appraisal of the test as a valid technique for detecting arsenic sulfide, see Yun Sik Nam, Sung-Ok Won, and Kang-Bong Lee, "Modern Scientific Evidence Pertaining to Criminal Investigations in the Chosun Dynasty Era (1392–1897 A.C.E.) in Korea," *Journal of Forensic Sciences* 59, no. 4 (2014): 974–7. I thank Dr. Jeffrey Jentzen for bringing this article to my attention.

[64] *Chongkan buzhu Xiyuan lu jizheng*, 3.44a–45b.

[65] Ken Alder, *The Lie Detectors: The History of an American Obsession* (New York: Free Press, 2007), 125–9 especially.

[66] *Fayi yuekan*, no. 11 (1934), 130.

washed with the soap-bean solution, discoloration caused by contact with poison would not.[67] Thus, much was at stake in the relative's washing of the needle.

In the end, even after repeated scrubbing, the discoloration on the needle did not disappear, a sure sign of poisoning and a refutation of the relative's claim that the deceased had been beaten to death. Nevertheless, with the accusation against the village head about to be put to rest, the relative again "provoked doubts among the majority of people" about the cause of death.[68] As in other cases involving disputed forensic evidence, county officials now contacted the Gansu High Court to facilitate a new examination of the questioned evidence. Rather than seeking an expert inspection clerk to re-examine the body, however, officials from this court sought a medical expert to carry out chemical testing (*huayan*) of the needle in order to establish conclusively whether or not the villager had been poisoned. After contacting Gansu Zhongshan Hospital, which claimed that it was unable to carry out such an examination, the court sent the needle to the Research Institute.

Cases such as this one reveal the distinctive engagements with laboratory science of officials who approached legal medicine from within their own tradition of forensic knowledge about bodies and things. Officials sent silver needles for testing because the colors that they observed did not match what was described in the *Washing Away of Wrongs* – an indication of the important role that this text continued to play in their understandings of forensic evidence. In one such case, sent to the medical school of Beiping University in early February 1936, the Hebei High Court was retrying a case of suspected poisoning.[69] An individual named Li Nairu had accused another, Song Ruiqi, of poisoning a Mrs. Song née Li, who had died almost immediately after taking a new medicinal decoction. The officials who initially investigated the case had used a 50-centimeter long silver rod to test for poison in the rectum of the corpse; it was not inserted into the mouth because the rapid decomposition of the body in the summer heat had made it difficult to manipulate the head. What prompted officials in this case to send the probe to the laboratory for testing was that there was an unexpected "bluish-red color" (*qinghong yanse*) that remained on the probe after it was removed from the body. The judges who had initially tried the case expressed their own doubts about whether these marks actually indicated that Mrs. Song née Li had been poisoned:

---

[67] *Lüliguan jiaozheng Xiyuan lu*, 3.17b. [68] *Fayi yuekan*, no. 11 (1934), 130.

[69] Beiping University Medical School Institute of Legal Medicine, Written appraisal report, February 5, 1936, BMA J29-3-608, 9–20; *New Medicine* (Xin yiyao zazhi) 4, no. 5, May 1936, 47–52.

When a poison enters the internal organs (*zangfu*), whichever part is affected by the poison will be the only one for which discoloration will appear [following insertion and removal of the testing needle]. There is no way that the poison would descend to the rectum and thus be found everywhere in the body. Moreover, when the deceased, Mrs. Song née Li, took the medicine, she quickly vomited it out, as described by her daughter Song Airu. It can thus be deduced that even if there had been poison, it could not have reached the rectum. Thus, there is no doubt that the bluish-red color that cannot be scrubbed away was simply the result of exposure to the rotting-corpse vapors (*fubai shiqi*).[70]

This explanation clearly raised enough questions among officials of the High Court to cause them to contact Lin Ji's forensic laboratory in Beiping with the request that it examine the probe. As the officials wrote in their request for testing: "We have the necessity of having appraised whether or not the bluish-red color on the silver probe is poison (*shifou duzhi*), and, if so, what kind of poison." This kind of query was similar to those in other cases in which silver needles were sent for laboratory testing. For example, in a case sent to the Research Institute in late June 1934 from officials of the procuracy of Nanzheng Local Court in Shaanxi, the officials' query revolved around a bluish-red color (*lanhong se*) that did not disappear from the needle after thorough scrubbing with soap pods and water. In this case, the officials' query was simply: "Are the colored marks on the needle poison (*zhenshang sehen shifou xi du*)? If they are, then what kind of poison?"[71]

These cases suggest a broader pattern: officials in China's courts and county offices were attempting to utilize the laboratory to resolve questions about evidence that had been collected on the basis of the *Washing Away of Wrongs*. The problem in such cases, at least from the perspective of Lin Ji, was that the evidence that officials sent for testing was useless in the regime of forensic proof associated with legal medicine. According to the basic principles that informed Lin Ji's own form of toxicological investigation, for example, evidence of poisoning could be obtained from chemical analysis of samples of the food or medicine that had been ingested or the contents of the stomach, analyzed in conjunction with clinical evidence regarding the condition of the poisoned individual and anatomical investigation of physical changes caused by a toxin in the body.[72] Any technique for the testing of poison not based on these

---

[70] Beiping University Medical School Institute of Legal Medicine, Written appraisal report, February 5, 1936, BMA J29-3-608, 17.

[71] *Fayi yuekan*, no. 11 (1934), 132–4.

[72] For Lin Ji's general discussion of poisoning, see Lin, *Fayixue gelun*, 169b–198a. Also see the brief discussion included in written appraisal reports that were returned to legal officials, i.e., Beiping University Medical School Institute of Legal Medicine, Written appraisal report, February 5, 1936, BMA J29-3-608, 18–19.

principles and, specifically, on understandings of physical evidence grounded in analytical chemistry, was not only incapable of producing authoritative toxicological evidence, but was erroneous if not ridiculous from the perspective of modern science. Officials who did not share these epistemological standards were likely to send evidence for testing, such as testing needles, that had no value in legal medicine. Moreover, such cases suggested that officials lacked an understanding of science-based forensic techniques. Thus, when officials of Nanyang Local Court in southwestern Henan sent the blackened fingernail of a suspected poisoning victim to Beiping for testing – livid fingernails, along with discolored skin, constituting an accepted sign of poisoning in the *Washing Away of Wrongs* – the subtitle that introduced the case when it was published in *Beiping Medical Journal* indicated the negative judgment of Lin Ji and his professional colleagues: "A correction of the errors made by Chinese legal officials in their use of medicine."[73]

Testing Silver Needles

The problem in such cases was not simply that the techniques that officials were using appeared to be mistaken from the perspective of modern legal medicine – a criticism that was leveled at various techniques contained in the *Washing Away of Wrongs*, as we have seen. The problem was, rather, that in sending physical evidence that had been collected on the basis of an alternative standard of proof in poisoning cases, officials had made it impossible for Lin Ji to provide a useful appraisal of the evidence. Disciplining officials so that their evidence collection techniques matched the norms of the distant laboratory thus became a practical exigency in the project of facilitating interprofessional cooperation with local courts. During the 1930s, the silver needle test for poisons became a particularly important target of these efforts. Through discussions in the appraisal reports sent back to officials, as well as in medical journals and other printed media, Lin Ji framed these cases as visible evidence of the unscientific practices that were still being used in Chinese courts. Specifically, Lin Ji argued, the use of the test revealed a basic error on the part of officials, who believed that the discolorations on the needles were evidence of poisoning instead of simply being the visible result of silver sulfide staining. As Lin Ji explained, the chemical reactions that caused these discolorations

[73] Beiping University Medical School Institute of Legal Medicine, Written appraisal report, April 3, 1936, BMA J29-3–608, 239–50; "Yige zhizhao de jianyan" [The inspection of a fingernail], *Beiping yikan* 4, no. 8 (1936): 55–9.

could be expected any time a silver object was exposed to chemical compounds such as hydrogen sulfide that were produced with the breakdown of proteins in a dead body. If such a probe were inserted into the throat of a "fresh corpse" (*xinxian shiti*), by contrast, the object would not tarnish because sulfur-containing compounds would not yet have been released. If the method were used to test for poisons in a decaying corpse and the probe came into contact with these substances, then a discoloration might appear on the probe regardless of whether the deceased person had been poisoned.[74]

This explanation reflected the basic assumption that modern analytical chemistry could provide the most authoritative understandings of the physical appearance and toxicological significance of this kind of evidence. Any explanation of the colors on the needles that was not based on this understanding of the chemical properties of these objects was simply untenable. Lin Ji's refutation of the silver needle test relied on laboratory experiments that demonstrated these points. Lin was not the first to subject the silver needle test to investigation in the laboratory as a means of demonstrating its questionable effectiveness as a test for poison. He might have been aware that during the late 1910s, faculty of his own alma mater, National Medical College in Beijing, had carried out a modest program of experiments. For example, a silver needle was inserted into the rectum of the unclaimed body of a prisoner who had died at Beijing No. 1 Prison, an early source of anatomical material for faculty and students at the school. Given that the prisoner had not died of poisoning but of illness (pulmonary consumption and dysentery), the dark discoloration that appeared on this needle only confirmed that the test lacked validity.[75] Lin Ji's own experiments surpassed these earlier tests in scope and visibility. They were described in *Monthly Bulletin of Legal Medicine* and other academic journals, the written appraisals that were returned to courts, and even the newspaper *Shenbao*, which published the full text of Lin Ji's critique of the test and description of the experiments.[76]

Lin Ji described a number of these experiments, for example, in a 1932 piece titled "Four kinds of small experiments in legal medicine," which appeared in one of the research journals published by the medical school

---

[74] Beiping University Medical School Institute of Legal Medicine, Written appraisal report, February 5, 1936, BMA J29-3-608, 12–4.

[75] Tang, "Xue fazheng de ren keyi budong xie yixue ma?," 298–9.

[76] Lin Ji, "Jianyan Xiyuan lu yinchai yandu fangfa buqie shiyong yijianshu" [An opinion regarding the unsuitability of the silver hairpin method of testing for poisons in the *Washing Away of Wrongs* used in forensic examinations], *Fayi yuekan*, no. 5 (1934): 53–6; SNB April 19, 1933, 10.

of Beiping University.[77] The article introduced Lin Ji's own practical experience with several areas of German and Japanese medico-legal innovation, including forensic applications of entomology to establish time since death and the use of ultraviolet rays to examine bodily fluids and other physical evidence. In the section of the article devoted to the silver needle test, Lin Ji began by critically evaluating passages from the *Washing Away of Wrongs* that described the use of the technique, arguing that these were inconsistent in failing to provide methods for testing all poisons mentioned in the text and too general in their failure to help officials differentiate between poisons. Lin Ji then described numerous experiments that he carried out in order to investigate interactions between silver objects, poisons, and chemical substances containing sulfur.

In the first experiment he inserted a silver needle into a sample of arsenous acid from a case that had been sent for appraisal by the Hebei High Court and boiled it for two hours, with no black marks appearing. After this test he switched to using a silver coin placed into an evaporating dish with arsenous acid crystals and tissue samples from decomposing intestines. After cooking the mixture for two hours, the coin did not turn black and Lin Ji found the same result after leaving it to soak for another day and night. A similar experiment performed with arsenic sulfide found that the coin immediately began to take on a darkish-yellow color, and after 15 minutes the entire coin had changed color. Lin Ji performed the same procedure again, but substituted realgar (*xionghuang*), an arsenic sulfide containing less sulfur content than the previous sample. That the coin only changed to a slight yellowish color proved that "silver turns black because of the sulfur and not because of the arsenic." This point was further demonstrated with experiments showing that silver objects did not change color when exposed to other toxins, but did when exposed to sulfur-containing substances such as human excrement and preserved duck's eggs.[78]

The significance of these experiments went beyond the basic chemistry lesson that silver tarnishes upon contact with sulfur, a fact that on its own held little relevance for the larger questions of state-building, professionalization, and forensic modernity that occupied Lin Ji. In a coda that followed his description of experiments refuting this test that appeared in his article, Lin Ji provided a broader interpretation of the history of forensic science in China that drew implicitly on the laboratory experiments that he had just described as an epistemological

[77] Lin, "Fayixue sizhong xiaoshiyan."   [78] Ibid., 309–12.

standard that could be used to refute the scientific validity of China's past traditions of forensic knowledge. The main narrative was one of China's early advances in forensics and later decline, leading to the unscientific practices of the present moment. Thus, Lin Ji noted that Song Ci's mid-thirteenth-century *Collected Writings on the Washing Away of Wrongs* was undoubtedly a major achievement, one that had been completed during a period in which "the science of the West was still in the dark ages, and our China had already made use of complex methods in criminal inspections." Nonetheless, China's forensic practices ultimately came to lag behind those of the West, developing highly erroneous and unscientific methods such as the silver needle test.[79] However admirable China's long history of forensics had been, at the start of the twentieth century such techniques could only be viewed as unusable or, as Lin Ji put it elsewhere, to have been the product of "the misunderstandings of ancient people."[80]

As these statements suggest, Lin Ji understood there to be a very clear distinction between forensic techniques that had been validated by science and those that had not, as well as between those associated with the globalizing modernity of his present moment and those that had developed under pre-modern conditions of seemingly imperfect natural knowledge. Yet, such assumptions about the fixed boundaries between scientific modernity and its Others were in many ways belied by the fact that forensic practice in 1930s China was characterized by a plurality of epistemologies, ways of legitimizing knowledge and expertise, and models of forensic practice. Judicial officials and their subordinate forensic personnel remained active in Beijing and elsewhere throughout the entire Republican period. Nor did medico-legal experts such as Lin Ji fundamentally challenge their professional jurisdiction over the forensic examination of corpses or the investigation of homicide. In fact, in extending the reach of legal medicine beyond Shanghai and Beiping, proponents of this scientific discipline had already made professional and epistemological compromises with the judicial officials whose evidence they tested in the laboratory. The fact that there already existed a judicial bureaucracy with standardized evidence collection procedures and long-distance networks of communication had made it possible for physical evidence from distant provinces to be examined in the laboratory in the first place.

---

[79] Lin, "Fayixue sizhong xiaoshiyan," 312.
[80] Beiping University Medical School Institute of Legal Medicine, Written appraisal report, February 5, 1936, BMA J29-3-608, 15.

Conclusion

In mid-September 1935, the Judicial Yuan of the Nationalist government convened a five-day long National Judicial Conference (Quanguo sifa huiyi) in Nanjing, the purpose of which was to assess the state of China's judiciary and consider proposals for its continued reform. The meeting was attended by more than 200 Nationalist government officials, central government and provincial judges and procurators, law school faculty, and others who were involved in some way with the judicial system. Sun Kuifang participated in the conference as director of the Research Institute of Legal Medicine; Lin Ji, who had left the directorship of this institution a number of months earlier, did not. Of almost 500 reform proposals that were deliberated on at the conference, 21 directly touched upon the forensic investigation of deaths, especially the training and deployment of the judiciary's forensic personnel.[81] With the exception of the proposal of Sun Kuifang and another by Wang Yongbin, head of the Ministry of Judicial Administration, all of the forensics-related proposals were made by procurators and judges of local and provincial courts, as well as the Supreme Court.

Sun Kuifang's proposal for reforming the judiciary's forensic examination practices had two parts. First, Sun suggested that some inspection clerks from each province should be sent to the Research Institute for a period of retraining, during which they would receive instruction regarding "which methods of the *Washing Away of Wrongs* accorded with science and thus were suitable for use, and which did not" and be given enough basic scientific knowledge to attain some degree of "common knowledge" (*changshi*) in legal medicine.[82] The second part of Sun's proposal was for the judiciary to revise – albeit not abandon – the standardized forms that inspection clerks were required to use when documenting the state of the corpse. Sun did not spell out the details of what the revised forms would look like in the proposal, aside from implying that they should avoid promoting the notion that one could decide cause of death on the basis of the location of a wound, and specifically whether it was on a "vital spot" or not.[83]

Sun's proposals accorded with the general tenor of the other forensics-related proposals at the conference. Many officials expressed dismay at the fact that old-style forensic examiners continued to handle

---

[81] *Quanguo sifa huiyi huibian, yian, di 4 zu*, 24–46.   [82] Ibid., *yian, di 4 zu*, 26–7.

[83] For a template of the revised forms that were ultimately drafted under Sun's directorship at the Research Institute, see Hu Qifei, "Xianxing yanduanshu pinglun ji xiugai zhi chuyi" [An assessment of the corpse examination forms currently in effect and a proposal for revision], *Fayixue jikan* 1, no. 3 (1936): 49–90.

most cases in local areas, that there were so few trained medico-legal experts, and that the salaries of China's forensic personnel were prohibitively low. The proposals that they made ranged from suggestions to expand the scope of the Research Institute and its work, to training intermediate tiers of medico-legal examiners in local areas, to retraining inspection clerks with a combination of legal medicine and the *Washing Away of Wrongs* – a strategy that was similar to the one that Sun himself proposed. There is little sense from the proposals that officials within the judiciary saw any contradiction between calling for experts in legal medicine to have greater influence, criticizing the shortcomings of the judiciary's forensic examiners and the *Washing Away of Wrongs*, and accepting that the latter still had a role to play in the present moment.[84] As a provisional measure, for example, the head of the procurator's office of the Supreme Court even proposed issuing a new edition of the *Washing Away of Wrongs*, revised so as to be more compatible with the forensic norms of legal medicine.[85] This would involve correcting or replacing the text's methods for testing for poisons and blood stains and examining bones, all of which purportedly originated in the "misunderstandings of ancient people" (*guren wuhui*), a phrasing that suggests the impact that Lin Ji and his colleagues were already having on the forensic norms of the judiciary. At the same time, of course, this and other proposals implicitly validated the judiciary's existing model of forensic practice, which relied on the use of a state-regulated body of forensic knowledge that could be used by non-expert judicial personnel in local courts and county judicial offices.

It is clear from these proceedings just how unrealistic Lin Ji's original vision of a physician-staffed national forensic infrastructure was during this period. While this ambitious goal reflected Lin Ji's desire for China to have the most advanced forensic examination system possible, it failed to take into account the institutional and financial challenges that faced the judiciary. When this particular vision of forensic modernity is viewed alongside judicial officials' own perspectives on the prospects for forensic reform, articulated forthrightly at the National Judicial Conference, it becomes clear that the idea that one could train a large group of medical specialists and employ them throughout the entire judicial infrastructure, top to bottom, was unrealistic from the start. In contrast, the prevailing sentiment at the conference, articulated by officials who were literally on the front lines of the administration of justice in the provinces, might be characterized as pragmatic: while it was desirable to establish a forensic examination system based on

---

[84] *Quanguo sifa huiyi huibian, di 4 zu*, 30.   [85] Ibid., *di 4 zu*, 37–8.

medico-legal expertise, the most immediate reforms should focus on improving the existing system. That Nationalist China's forensic examination system was, in a sense, already *modern* – that is, that it supported the operation of modern courts and procuracies throughout the country and even facilitated the integration of medico-legal laboratories into this existing judicial infrastructure – was a point left unsaid by the participants of the conference, but a reality that would have been difficult to deny.

Conclusion: A History of Forensic Modernity

In mid-August 1942, Yu Yuan and Fu Changlin proposed to the Beijing procuracy that it train a new group of inspection clerks through a year-long training program in forensic examination.[1] By this point, Beijing had been under Japanese occupation for five years, a period that saw rising inflation and economic uncertainty, an evolving administration of Japanese and Chinese military and political authorities, and the implementation of a repressive regime of policing and social control in the city.[2] In the investigation of homicide, there were many elements of continuity with earlier decades. Much as before, police and judicial officials continued to investigate suspicious deaths and homicides, at times with the assistance of a new complement of Japanese faculty and staff who succeeded Lin Ji in the department of legal medicine of the medical school of Beijing University.[3] The procuracy continued to rely on about a dozen inspection clerks to examine bodies, and also employed an individual named Zhang Huazhong as the office's medico-legal physician (*fayishi*), a position that did not exist in Beijing a half-decade earlier.[4] It is likely that Yu and the other forensic personnel foresaw their impending retirement. By the early 1940s, some of them – Yu Yuan and Song Ze, for example – had been serving in this position for decades.

Following approval by the procuracy, the training program began in September 1942. The new trainees, of which there were initially four, were all related to existing staff of this office. These included Yu Yuan's

---

[1] Petition of Yu Yuan and Fu Changlin, August 15, 1942, BMA J174-2-52, 1–4.

[2] Sophia Lee, "Education in Wartime Beijing: 1937–1945" (PhD dissertation, University of Michigan, 1996), 26–81.

[3] For more on the Beijing University Medical School legal medicine department during this period, see *Tokyo teikoku daigaku hōigaku kyōshitsu gojūsannen shi* [Fifty-three year history of the Tokyo Imperial University Department of Legal Medicine] (Tokyo: Tokyo teikoku daigaku igakubu hōigaku kyōshitsu, 1943), 564–5; *Guoli Beijing daxue yixueyuan zhijiaoyuan xuesheng tongxinlu* [Directory of administrative and teaching staff and students of National Beijing University Medical School] (Beijing: National Beijing University Medical School, 1940), 10, 15, 18.

[4] "Register of staff of Beijing Local Court Procurator's Office," January 1941, BMA J174-2-239, 36–7.

16 *sui* great-grandson Yu Zhiwen, a 16 *sui* member of the Song clan and nephew of inspection clerk Song Chunyi who was named Song Zhaoyi, and the 20 *sui* younger brother of another Beijing examiner named Qian Songtao. The fourth trainee, 16 *sui* Jin Hengxi, was the son of a member of the procuracy's office staff, and unrelated to existing forensic personnel. By November 1942, the group had acquired three more students, including an 18 *sui* female student named Liu Xiuzhen, and Wu Jingbo, the 24 *sui* relative of another inspection clerk in the procuracy. Another female student, 24 *sui* Jiang Fengyi, subsequently joined the class.[5]

What is particularly interesting about this instance of training is the way in which legal medicine and the *Washing Away of Wrongs* were integrated in the instructional curriculum. Procurator Ming Yan, who had overseen the examination of Zhang Shulin's dismembered body in the East Station six and a half years earlier, taught passages from the *Washing Away of Wrongs*. Zhang Huazhong instructed the trainees in a unit titled "main points of legal medicine" (*fayixue gangyao*), and Yu Yuan and Fu Changlin instructed the students during the remaining time.[6] The students were taught new understandings of the structure of the skeleton (now drawn from Western anatomy and not the *Washing Away of Wrongs*), the causes and manifestations of rigor mortis and post-mortem lividity, and the physical effects of external trauma on the body, now understood in terms of blood vessels, tissues (*zuzhi*), and other structures defined by anatomical science. Trainees were not tested on methods for testing blood stains and poisons from the *Washing Away of Wrongs*, a clear departure from early Republican training curricula and likely a reflection of the criticisms that had been launched against these methods over previous decades. Nonetheless, the core of the forensic knowledge learned by the trainees remained the *Washing Away of Wrongs* and the officially endorsed body examination methods that even in the 1940s continued to be used in homicide cases. The old conception of the "vital" and "non-vital" points of the body around which these methods were organized was one of the subjects on which the trainees were tested.[7]

This training program provides both a fitting endpoint for the history of Chinese forensics explored in this book and a case that exemplifies some of the larger themes of professionalization, science, and modernity that have

---

[5] Biographical information can be gleaned from lists of enrollees, guarantors' certificates, and other documents contained in BMA J174-2-52.

[6] Beijing Local Court Procurator's Office to Hebei High Court Procurator's Office, August 22, 1942, BMA J174-2-52, 12–14.

[7] The students' testing papers, which show their grasp of these subjects, are included in BMA J174-2-52.

been central to this story. This book has focused, at one level, on the history of formal institutions of professional science – academic departments, research institutes, professional associations, and the related activities of profession-building, publicity, and credentialing that these institutions and their experts pushed forward. Likewise, the activities of individuals such as Tang Erhe, Xu Songming, Lin Ji, and Sun Kuifang have been particularly visible elements of the story. Lin Ji himself is often referred to as the "founder" (*dianji ren*) of modern legal medicine in China, a judgment that reflects both the breadth and persistence of his discipline-building activities and the important role that his students have played in building legal medicine after 1949. From such a profession-centered perspective, the individual actions of Lin Ji and other medico-legal experts, and the institutions that they worked to build, stand out as particularly important elements of the history of forensics in early twentieth-century China.

It is important to acknowledge, of course, that this discipline-building project was one that saw halting and uneven successes during this period. Major challenges included political fragmentation at national and regional levels, the chronic underfunding of judicial institutions, shortages of trained physicians of Western medicine, and the social displacements and physical destruction caused by decades of intermittent warfare.[8] After 1937, the protracted war with Japan that was waged within China itself further disrupted the development of medico-legal institutions, most visibly with the destruction of the Research Institute of Legal Medicine during the Japanese bombardment of Shanghai and its surrounding areas.[9] A number of personnel associated with Nanjing-decade legal medicine, including both Sun Kuifang and Lin Ji, left Japanese-occupied areas of coastal China and followed the Nationalists into the interior. For Lin Ji, this journey wound from Beijing to Shaanxi, where he served for a time on the faculty of the Northwestern Associated University (Xibei lianhe daxue), and then to Chongqing, where he assisted once again in the drafting of medical laws under the Nationalist government while pursuing a new series of forensic training programs through the medical school of Central University.[10]

[8] For an overview of legal medicine in early twentieth-century China that largely adopts a profession-focused narrative of promising institutional developments in the midst of a period of warlordism and reactionary social politics, see Jia, "Xinhai geming yihou de zhongguo fayixue."

[9] Chen Kangyi, who worked at the Research Institute after graduating in its first class of researchers in 1934, claimed that the facility was destroyed, leaving nothing more than "scorched earth" (*jiaotu*). Chen Kangyi, "Zhongguo fayixue shi" [A history of legal medicine in China], *Yishi zazhi* 4, no.1 (1952): 6.

[10] Chen, "Zhongguo fayixue shi," 6–7; Huang, *Fayi qingtian*, 101–8; Huang, *Zhongguo jinxiandai fayixue fazhan shi*, 74–5.

From such a perspective, which focuses on the actions of medico-legal experts and the professional structures that they created, the Beijing procuracy's early 1940s training program was a development of ambiguous meaning. While the prominence of legal medicine in the instructional curriculum certainly suggests the influence that this discipline had already attained in Beijing and elsewhere, it is important to remember that the point of the training was to maintain the capacity of the judiciary's own personnel to carry out forensic examinations, not to consolidate professional physicians' exclusive authority over this area of judicial practice. The peripheral forensic role played by professional physicians is another theme that has emerged in this account of early twentieth-century Chinese forensics, and one that would seem also to be reflected in the role that Zhang Huazhong played in Beijing as a medico-legal expert who supplemented rather than replaced the judiciary's own forensic personnel.

As this instance of training suggests, however, the persistent institutional weakness of professional legal medicine should not be taken as evidence that medico-legal science was irrelevant to the judiciary. Indeed, some of the earliest proponents of legal medicine in China were the Qing officials who attempted to use this new field of scientific knowledge to improve the existing system of *wuzuo*; in the process, they hoped to bring the Qing empire's forensic practices more closely into line with those of Japan. In this and other instances, including the training program in Beijing, one thus finds a history of "legal medicine" in China that parallels medical professionals' attempts to promote the discipline, yet in which the main agents are not professional physicians or medico-legal experts but, rather, judicial officials. The agency of individual medical reformers – and, by implication, the successes or failures of particular professional or disciplinary programs – is less important to this story than are the broader forces unleashed under the new historical conditions of the early twentieth century that invested legal medicine, and other fields of scientific and professional knowledge, with authority. These included, in broad strokes, Euro-Japanese imperialism, extraterritoriality as an impetus for judicial reform, and the drive to build a strong state comparable to those of industrialized countries. It was within these broader contexts that legal medicine gained authority and legitimacy in China as a body of modern knowledge that was relevant to the state and had value even for judicial officials who continued to rely on the *Washing Away of Wrongs* in practice.

The training program carried out in Beijing in 1942–3 exemplifies this world of judicial officials' vernacular science. As we have seen, the judiciary's forensic personnel might be selectively trained in new Western-anatomical conceptions of the body, much as the *Washing Away of Wrongs* could be modified to accord with new epistemological standards associated with legal medicine. Likewise, officials might seek laboratory testing to address forensic questions raised by the *Washing Away of Wrongs*, a practice that implicitly demonstrated their acceptance of the authority of the laboratory. These uses of modern Western science, potentially ambivalent from the perspective of medico-legal professionals themselves, were at once accepting of epistemological pluralism and pragmatic. Republican judicial officials' flexible orientation toward forensic knowledge was perfectly demonstrated by the officials of Hengshui county who in 1929 sought outside assistance in the case of Zhang Wenhuan because, in their words, both "the *Washing Away of Wrongs* of olden times and the legal medicine books of the present" (*xi zhi Xiyuan lu yu jin zhi fayixue dengshu*) lacked methods for confirming that the deceased had been poisoned by golden pills.[11] Thus, there was little sense of absolute epistemological hierarchy: it was more important to identify techniques that worked in practice and to use them.

That medical professionals were less likely to acknowledge or accept value in epistemological syncretism of this kind reflects the importance that they placed on drawing boundaries based on the epistemological authority of particular forensic examination techniques (that is, whether they were grounded in "science" or not) as well as their historical temporality (whether they were "modern" or not). The practical benefit of making such distinctions was that doing so made it possible to put forward the argument that absolute distinctions existed between the different capacities of legal medicine and the *Washing Away of Wrongs* to have value and be effective under the new conditions of the modern age, a claim that was inseparable from physicians' pursuit of professional authority.[12] Thus, from the perspective of China's first medico-legal professionals, the forensic hybridities examined here might be interpreted, at best, as half-measures toward the fullest medicalization of Chinese forensic practices, accomplished when the forensic work of judicial officials and inspection clerks was replaced with that of physicians of Western scientific medicine. At worst, such practices might simply demonstrate judicial officials' "errors" in their use of legal medicine,

[11] Procurator's Office of Hebei High Court to Chief Procurator of Beiping Local Court, October 30, 1929, BMA J174-2-279, 29.

[12] For a broader perspective on the role that boundary-drawing of this kind has played in the rise of industrialized societies (and their particular forms of science, technology, political organization, and ideology) as well as in the process through which the people who belong to these societies have elevated themselves, discursively, over purportedly non-modern societies and cultures, see Bruno Latour, *We Have Never Been Modern*, translated by Catherine Porter (Cambridge: Harvard University Press, 2002).

much as we saw in the case of officials' practice of sending silver testing probes and other "unsanctioned" objects for laboratory testing.

Of course, one cannot easily separate the profession-focused history of legal medicine from that of the judiciary's own appropriation of medico-legal science. Indeed, it would be difficult to tell the story of Lin Ji, the Research Institute of Legal Medicine, and so on, without addressing also the role that judicial officials played in the day-to-day operations of professional legal medicine and, more broadly, the judicially organized field of forensic practice over which medico-legal experts struggled. Likewise, it would be difficult to write a history of inspection clerks and the *Washing Away of Wrongs* without addressing the points at which this regime of forensic knowledge and practice encountered legal medicine and its new expectations for science-based professional expertise.

For these reasons, the story told here might be interpreted productively as a case of "coevolutionary history," a concept that Sean Lei has used to characterize the relationship between the histories of Chinese medicine, Western scientific medicine, and the modern state in early twentieth-century China.[13] Lei has argued that histories of Chinese and Western medicine cannot be isolated from each other and, moreover, that our understanding of the history of Chinese medicine has tremendous bearing on how we understand the modern Chinese state's broader engagements with medicine and healthcare. A similar point could be made for the "reciprocal interactions" (to borrow Lei's words) that defined the history of forensics in early twentieth-century China. The old forensic institutions of the Qing state were transformed through engagements with new concepts, institutions, techniques, and possibilities associated with Western scientific medicine and the laboratory-based knowledge on which it was based. In establishing an institutional niche for themselves in the Republican judiciary, proponents of legal medicine had to forge their own connections with judicial officials. By the mid-1930s, the professional jurisdiction that medico-legal experts had attained was itself a result of the decisions and actions of local officials, who sent physical evidence to the laboratory on the basis of their own practices of homicide investigation and in response to local contestations over evidence in particular cases. The history of legal medicine in early twentieth-century China thus cannot be narrated without attention to the negotiation and interplay that unfolded between these different forms of forensic knowledge and practice.

While legal medicine did emerge in these ways as an unavoidable force – discursively and institutionally – in early twentieth-century

---

[13] Lei, *Neither Donkey nor Horse*, 7–10.

China's forensic scene, it is important that we do not assume teleologi-
cally that the forensic institutions and practices of the Qing had to
undergo a corresponding and inevitable decline. Indeed, the story told
here has been one of radical contingency and unexpected outcomes of
modernity that made the *Washing Away of Wrongs* and the forensic
practices associated with it an important element of forensic, judicial,
and urban modernity in Republican China. I have argued that this regime
of institutionalized knowledge maintained authority and legitimacy
because of a series of changes that must be located *within* the modern
itself: the reform of the judiciary created a strong, institutionally protected
position for inspection clerks; the creation of a new kind of police force in
Beijing created a demand for their forensic knowledge and skills; new
journalistic practices legitimized this work through newspaper representations of urban policing and everyday life. In the process, inspection
clerks came to play an important role in the establishment of modern
institutions and new patterns of state–society relations. They consolidated procurators' control over physical evidence, facilitated the creation
of a particular kind of policed urban order in Beijing, and even assisted
medico-legal laboratories in the collection of physical evidence.
Inspection clerks were not recognizable as modern "professionals" –
they did not derive authority from the patterns of training, credentialing,
or public engagement that supported the new expert-claims of physicians
of Western scientific medicine and other groups that claimed professional
identity. Nonetheless, the particular way of organizing forensic knowledge and expertise with which inspection clerks were associated provided
special utility to a state undergoing the throes of modernization.

The urban has been an important site at which to understand the modern
transformation of Chinese forensics as it has been examined in this book.
One might expect, quite understandably, that the city would be an important site for investigating interactions between medical professionals and
the state given the important role that the shifting economic patterns,
infrastructure, and social relationships of the urban context played in the
rise of this and other professional groups.[14] That modern cities have been
important sites for the development of scientific institutions as well as the
practical application of scientific knowledge and expertise in public hygiene
and other areas of governance is borne out by the case of Beijing, a city that
contained important and municipally involved schools of modern higher
education such as National Medical College and PUMC.[15] Early

[14] Xu, *Chinese Professionals and the Republican State*, 23–49.
[15] For general perspectives on science and the urban context, see Sven Dierig,
Jens Lachmund, and J. Andrew Mendelsohn, "Introduction: Toward an Urban
History of Science," *Osiris*, 2nd Series, Vol. 18, Science and the City (2003): 1–19.

twentieth-century Beijing also boasted a significant concentration of modern judicial and policing institutions, as well as an active legal profession.[16] My approach, however, has not been to seek a smooth or coherent convergence of modern institutions and practices in the urban context, but, rather, to focus on the city as a site at which myriad modern forces – social, institutional, intellectual, and technological – were brought into play simultaneously, without coordination, and thus were productive of new social patterns, outcomes, and risks in their unpredictable interactions.[17]

This book has thus sought modernity in the unintended destruction caused by streetcars and automobiles in populated urban spaces, the emergent interactions between policing and journalism in the creation of a policed urban order, the newspapers' vacillating support and derision of medical institutions and professionals, and the complex and unwieldy set of claims that came to surround the urban dead. That the *Washing Away of Wrongs* and the larger world of Qing forensic practice were put into the service of modern policing and homicide investigation must be understood in this context as one of many examples of "recycling" – to return to Madeleine Yue Dong's important metaphor – that shaped everyday life in Republican Beijing.[18] Acknowledging without qualification the *modernity* of this at-once old and new forensic regime and the broader set of state–society interactions and cultural meanings that came to surround it allows for a more complex understanding of the different forms that the modern has taken as it has emerged in different times and places. From this perspective, the *Washing Away of Wrongs* can truly become part of the global history of forensic science, not as a seemingly eccentric accomplishment of pre-modern China that anticipated the rise of Western legal medicine, but, rather, as the subject of its own history of forensic modernity, one that highlights the complex historical genealogies of twentieth-century forensic science as it has emerged from the negotiation of old and new institutions and practices and the unpredictable interplay of law, science, and politics.[19]

---

[16] Ng, *Legal Transplantation in Early Twentieth-Century China*.

[17] For a critique of reductionist definitions of modernity and a powerful assertion of the importance of acknowledging and empirically investigating "the plurality of the processes that constitute modernity by their historical combination," see Sudipta Kaviraj, "Modernity and Politics in India," *Daedalus* 129, no. 1, Multiple Modernities (2000): 139–40.

[18] Dong, *Republican Beijing*.

[19] On this point, see especially Ian Burney's account of the coroner's inquest in late-nineteenth- and early twentieth-century Britain. As Burney argues, this old institution became implicated in a new politics of professional medical expertise that challenged its legitimacy even as it invested it with practical value as a potential vehicle for the expansion of medical authority and political meaning as a symbol of democratic participation. Burney, *Bodies of Evidence*.

Writing on the importance of understanding the modern condition as a set of underdetermined possibilities that are shaped by specificities of time and place, Carol Gluck has written that "[history], in short, offers no abstract model of the modern, only embedded real modernities produced by creative blending that never reaches an end."[20] Modernity thus emerges as the specific product of contingent interactions that take place between globalizing political, economic, technological, and social forces and the "preexisting conditions," also of course in motion, that pertain to particular times, places, and moments. The particular forensic modernity that pertained to homicide investigation in early twentieth-century Beijing was shaped at once by the new and powerful transnational forces that have made modern science a form of knowledge that is understood to be epistemologically authoritative and politically necessary across the globe; yet, it was also shaped by an older configuration of institutions, knowledge, and power that had supported the legal system of a powerful early modern empire and one that had, in its own ways, made compelling claims to expansive territoriality, administrative effectiveness, and epistemological authority. If we view modernity as a series of discrete moments of interaction and negotiation that are always unfolding in unpredictable ways, the particular moment of modernity that this book has examined must be located quite specifically in the immediate aftermath of the collapse of the Qing, the new possibilities and challenges afforded by the forces of destruction and reform that followed in its wake, and the ways in which cases of suspicious death and homicide became an arena for renegotiating the social order under these new conditions.

---

[20] Carol Gluck, "The End of Elsewhere: Writing Modernity Now," *American Historical Review* 116, no. 3 (2011): 686.

Glossary

Ai Shiguang　艾實光
anlian　諳練
Ban'an yaolüe　辦案要略
banguo cheng'an　辦過成案
baxian zhuo　八仙桌
Beiping difang fayuan　北平地方法院
Beiping yikan　北平醫刊
Beiping zhentan an　北平偵探案
Bi Xiugu　畢秀姑
biansi　變死
Bingyin yixue she　丙寅醫學社
binyiguan　殯儀館
bu zhiming　不致命
Cai Jiahui　蔡嘉惠
Cai Yuanpei　蔡元培
caipan yixue　裁判醫學
can　慘
can bu ren kan　慘不忍看
changshi　常識
Che Qingyun　車慶雲
Chen Kangyi　陳康頤
Chen Kuitang　陳奎棠
Chen Yuan　陳垣
Chen Zhiqian　陳志潛
cheng'an　成案
Chenzhong bao　晨鐘報
chubin zhizhao　出殯執照
daobi　倒斃
daobi pinmin　倒斃貧民
daode　道德
daode renge　道德人格

daotu daobi   道途倒斃

daowo   倒臥

Daxue yuan   大學院

dianji ren   奠基人

dibao   地保

Ding Fubao   丁福保

Ding Yulin   丁玉林

Dongling   東陵

Du Keming   杜克明

Duan Qirui   段祺瑞

duomin   墮民

ergen gu   耳根骨

Faguan xunliansuo   法官訓練所

Fang Yiji   方頤積

Fayi yanjiusuo   法醫研究所

Fayi yuekan   法醫月刊

fayi zhishi   法醫知識

fayishi   法醫師

fayixue   法醫學

fayixue gangyao   法醫學綱要

Fayixue jikan   法醫學季刊

fayixuejia   法医学家

fazhong   發塚

fei lüshi   非律師

fei zhiming   非致命

Feng Xiangguang   馮祥光

Feng Yuxiang   馮玉祥

fengsu   風俗

Fu Changlin   傅長林

Fu Changling   傅長齡

Fu Shun   傅順

fubai shiqi   腐敗屍氣

Ge Pinlian   葛品連

ge yamen jingxi wuzuo   各衙門經習仵作

Gong Shucang   龔漱滄

gongmu   公墓

Gu Nanqun   顧南群

gudai fayixue   古代法医学

guofa   國法

guomin geming   國民革命

guoqi   國旗

guren wuhui   古人誤會

He Kang　賀康
Huang Qinglan　黃慶瀾
huayan　化驗
Hui Hong　惠洪
huiguan　會館
huiqi　穢氣
hunsang　婚喪
Ishikawa Kiyotada　石川清忠
Jia Kui　賈魁
jianding　鑑定
jianding ren　鑑定人
jianding shu　鑑定書
Jiang Fengyi　姜逢怡
Jianghai　殭骸
jianmin　賤民
jianyan　檢驗
Jianyan jizheng　檢驗集證
jianyan li　檢驗吏
jianyan yuan　檢驗員
jianzheng ren　見證人
jiaotu　焦土
jichu　基礎
Jin Dasheng　金達生
Jin Hengxi　金恆喜
jindan　金丹
Jing Tai　景泰
jing yu zhengyan　精於蒸驗
Jingbao　京報
Jingbao tuhua zhoukan　京報圖畫週刊
Jingguan gaodeng xuexiao　警官高等學校
Jingshi difang jianchating　京師地方檢察廳
Jingshi fazheng xuetang　京師法政學堂
Jingshi jingchating　京師警察廳
jingyan　經驗
jinshi zhengjian　近時蒸檢
Kawashima Naniwa　川島浪速
kedian bingwang　客店病亡
kexue　科學
lanhong se　藍紅色
laoli　老吏
laolian zhi jianyan li　老練之檢驗吏
Li Maolin　李茂林

Li Nairu　李乃如
Li Shizhen　李時珍
Liang Qichao　梁啟超
Liang Zhenwen　梁振文
Liangbi　良弼
liangmin　良民
lieshi　烈士
Liexing dupin jiechusuo　烈性毒品戒除所
Lin Benyuan　林本元
Lin Zexu　林則徐
Lin Zhen'gang　林振綱
lingchi　凌遲
Liu Chongyou　劉崇佑
Liu Guangju　劉廣聚
Liu Lianbin　劉廉彬
Liu Ma　劉媽
Liu Qipeng　劉啟鵬
Liu Ruiheng　劉瑞恆
Liu Xiuzhen　劉秀貞
Liumin yangbingsuo　流民養病所
liyou　理由
Lu Xun　魯迅
Lüliguan jiaozheng Xiyuan lu　律例館校正洗冤錄
Luo Zhenqiu　羅鎮球
Ma Yulin　馬玉林
maiguo zhengfu　賣國政府
Ming Yan　明炎
Minguo yixue zazhi　民國醫學雜誌
mingyu jishu zhuanyuan　名譽技術專員
Moucai haiming an zhi yanming,
　　zhiming shang zai taiyangxue　謀財害命案之驗明, 致命傷在太陽穴
Nanba Mokusaburō　南波杢三郎
Nan'gang yidi　南崗義地
Nanyang yixue zhuanmen xuexiao　南洋醫學專門學校
neidi　內地
Puyi　溥儀
Qian Songtao　錢松濤
Qian Xilin　錢錫霖
qigai　乞丐
Qigai shourongsuo　乞丐收容所
qinghei　青黑
Quanguo sifa huiyi　全國司法會議

Quanti xinlun　全體新論
Quanyin　全印
renkou　人口
renqing　人情
Ronglu　榮祿
Satsujin kagakuteki sōsahō　殺人科學的搜查法
shaojiu　燒酒
shashang　殺傷
Shen Baozhen　沈葆楨
Shen Ruihong　沈睿洪
Shen Ruilin　沈瑞麟
Shen Zhiqi　沈之奇
shenghuo fanying　生活反應
Shenti fafu shou zhi fumu, bu gan huishang,
　xiao zhi shi ye　身體髮膚受之父母, 不敢毀傷, 孝之始也
shi　氏
Shi gong'an　施公案
Shi Jianqiao　施劍翹
Shi Shilun　施世綸
Shibao　實報
shifou duzhi　是否毒質
Shijing　詩經
shimin　市民
shirong　市容
shishi　事實
Shishi xinbao　時事新報
shiyanpin　試驗品
Shiyong fayixue　實用法醫學
si yu feiming　死於非命
sifa　司法
Sifa chucaiguan　司法儲才館
Song Chunyi　宋純義
Song Ci　宋慈
Song Duo　宋鐸
Song Minghui　宋明惠
Song Qixing　宋啟興
Song Ruiqi　宋瑞祺
Song Yueshan　宋月山
Song Ze　宋澤
Song Zhaoyi　宋昭沂
Su Baojie　粟保傑
Su Junchen　粟俊臣

Sun Chuanfang   孫傳芳
Sun Dianying   孫殿英
Sun Kuifang   孫逵方
Sun Shoushan   孫壽山
taimai zhizhao   抬埋執照
taiyangxue   太陽穴
Takayama Masao   高山正雄
Tan Jijian   譚季緘
Tang Erhe   湯爾和
Tang Hualong   湯化龍
Tian Chou   田疇
Tingsong qieyao   聽訟挈要
tongji diaochayuan   統計調查員
touruan   透軟
Wang Chichang   王熾昌
Wang Huayi   王化一
Wang Ligeng   王歷耕
Wang Mingde   王明德
Wang Weisan   王慰三
Wang Xichi   王錫熾
Wang Yongbin   王用賓
Wang You   王佑
Wang Youhuai   王又槐
Wang Yu   王玉
Wang Zhuyu   王柱宇
weisheng   衛生
wenpo   穩婆
wu   忤
Wu Jingbo   武靜波
Wu Kaicheng   吳闓澄
Wu Peifu   吳佩孚
Wu Zehan   吳則韓
Wusheng an   五聖庵
wuzuo   仵作
xi zhi Xiyuan lu yu jin
    zhi fayixue dengshu   昔之洗冤錄與今之法醫學等書
Xia Leibo   夏纍伯
Xia Quanyin   夏全印
xiandai fayixue   现代法医学
xiangyan   相驗
Xiaojing   孝經
xiati   下體

Xibei lianhe daxue　西北聯合大學

Xin yiyao zazhi　新醫藥雜誌

xingke shuli　刑科書吏

Xingshi susong tiaoli　刑事訴訟條例

xingzheng yamen　行政衙門

xingzheng yixue　行政醫學

xinxian shiti　新鮮屍體

xionghuang　雄黃

xiren　西人

xiyan　洗罨

Xiyuan jilu　洗冤集錄

Xiyuan lu　洗冤錄

Xiyuan lu xiangyi　洗冤錄詳義

Xu Lian　許槤

Xu Shichang　徐世昌

Xu Shijin　許世瑾

Xu Songming　徐誦明

Xuan Axiang　宣阿香

Xuan Yaojun　宣耀君

Xue Dubi　薛篤弼

xueli　學理

Yan Fuqing　顏福慶

Yang Hongtong　楊鴻通

Yang Naiwu　楊乃武

yangbang　殃榜

yangshu　殃書

Yao Zongxian　姚宗賢

yaohai chu　要害處

Ye Zaijun　葉在均

yiban yishi　一般醫師

yinyang sheng　陰陽生

yinyang xiansheng　陰陽先生

yiqie putong anjian　一切普通案件

yishi　醫師

Yiyao pinglun　醫藥評論

yizhong zhuanmen kexue　一種專門科學

yongyi　庸醫

you shuoming xi ju xueli shishi　右說明係據學理事實

Yu Delin　俞德林

Yu Depei　俞德霈

Yu Tao　俞濤

Yu Yuan　俞源

Yu Yunxiu　余雲岫
Yu Zhiwen　俞治文
Yuan Shikai　袁世凱
yuehu　樂戶
yushen　豫審
zangfu　臟腑
zaojia　皂莢
zaojiao　皂角
Zengfu　增福
Zhang Gongliang　张功良
Zhang Huazhong　張化中
Zhang Jiayi　張嘉懿
Zhang Ruilin　張瑞林
Zhang Shulin　張樹林
Zhang Wenhuan　張文煥
Zhang Yuanjie　張元節
Zhang Zongchang　張宗昌
Zhang Zuolin　張作霖
Zhao Fuhai　趙福海
Zhao Fukui　趙福魁
Zhaogong ci　趙公祠
zhengjian　蒸檢
zhenshang sehen shifou xi du　針上色痕是否係毒
zhentan xiaoshuo　偵探小說
zhiming　致命
zhiming chu　致命處
zhiming shang　致命傷
zhiming zhi suo　致命之所
Zhiwen xuehui　指紋學會
Zhiwen zazhi　指紋雜誌
Zhonghua minguo yiyaoxue hui　中華民國醫藥學會
Zhongyang zhengzhi huiyi　中央政治會議
Zhou Enlai　周恩來
Zhu Jiabao　朱家寶
Zhu Shixun　朱世勳
zhuidao hui　追悼會
ziyou zhiye zhe　自由職業者
Zizhi shiwu jianlichu　自治事務監理處
zuzhi　組織

Bibliography

Abbreviations

BMA	Beijing Municipal Archives
CB	*Chenbao* (Morning Post)
FHA	First Historical Archives of China
RAC	Rockefeller Archive Center
SB	*Shibao* (Truth Post)
SJRB	*Shijie ribao* (World Daily)
SNB	*Shenbao*

Abbott, Andrew. *The System of Professions: An Essay on the Division of Expert Labor.* Chicago: The University of Chicago Press, 1988.

Adas, Michael. *Machines as the Measure of Men: Science, Technology, and Ideologies of Western Dominance.* Ithaca: Cornell University Press, 1989.

Åhrén, Eva. *Death, Modernity, and the Body: Sweden 1870–1940.* Rochester: University of Rochester Press, 2009.

Aisin-Gioro Pu Yi. *From Emperor to Citizen – The Autobiography of Aisin-Gioro Pu Yi. Translated by W.J.F Jenner.* Peking: Foreign Languages Press, 1979.

Alder, Ken. *The Lie Detectors: The History of an American Obsession.* New York: Free Press, 2007.

Alford, William P. "Of Arsenic and Old Laws: Looking Anew at Criminal Justice in Late Imperial China." *California Law Review* 72, no. 6 (1984): 1180–256.

Altehenger, Jennifer. "Simplified Legal Knowledge in the Early PRC: Explaining and Publishing the Marriage Law." In *Chinese Law: Knowledge, Practice and Transformation, 1530s to 1950s,* edited by Li Chen and Madeleine Zelin. Leiden: Brill, 2015.

Ambage, Norman and Michael Clark. "Unbuilt Bloomsbury: Medico-Legal Institutes and Forensic Science Laboratories in England between the Wars." In *Legal Medicine in History,* edited by Michael Clark and Catherine Crawford. Cambridge: Cambridge University Press, 1994.

American Association for the Advancement of Science. *Reports on the Use of Expert Testimony in Court Proceedings in Foreign Countries.* Washington: Press of Byron S. Adams, 1918.

Andrews, Bridie. *The Making of Modern Chinese Medicine, 1850–1960.* Vancouver: UBC Press, 2014.

Ariès, Philippe. *The Hour of Our Death*. Translated by Helen Weaver. New York: Alfred A. Knopf, 1981.

Arnold, David. *Science, Technology and Medicine in Colonial India*. Cambridge: Cambridge University Press, 2004.

Asen, Daniel. "The Only Options?: 'Experience' and 'Theory' in Debates over Forensic Knowledge and Expertise in Early Twentieth-century China." In *Historical Epistemology and the Making of Modern Chinese Medicine*, edited by Howard Chiang. Manchester: Manchester University Press, 2015.

"Vital Spots, Mortal Wounds, and Forensic Practice: Finding Cause of Death in Nineteenth-Century China." *East Asian Science, Technology and Society: An International Journal* 3, vol. 4 (2009): 453–74.

Asen, Daniel and David Luesink. "Globalizing Biomedicine through Sino-Japanese Networks: The Case of National Medical College, Beijing, 1912–1937." In *China and the Globalization of Biomedicine*, edited by David Luesink, William H. Schneider, and Zhang Daqing. Under review.

Ash, Eric H. *Power, Knowledge, and Expertise in Elizabethan England*. Baltimore: The Johns Hopkins University Press, 2004.

"Banfa yanduanshu jianduanshu bing shangdan geshi ling" [An order on the promulgation of the corpse examination form, skeletal examination form, and wound list]. *Sifa ligui bubian* [Supplementary Collection of Judicial Regulations]. Beijing: Sifa gongbao faxingsuo, 1919, 238–61.

Becker, Elisa M. *Medicine, Law, and the State in Imperial Russia*. Budapest: Central European University Press, 2011.

"Beiping dongchezhan xiangshi an zhi jianding" [The appraisal in the Beiping East Station box-corpse case]. *Beiping yikan* 4, no. 7 (1936): 59–63.

Beiping shi weishengju dier weishengqu shiwusuo nianbao [Annual report of the Second Health District Station of the Beiping Bureau of Hygiene]. Beiping: Beiping shi weishengju dier weishengqu shiwusuo, 1935.

Beiping shi zhengfu weishengchu yewu baogao [Work report of the Beiping Municipal Government Hygiene Office]. Beiping: Beiping shi zhengfu weishengju, 1934.

Belsky, Richard. *Localities at the Center: Native Place, Space, and Power in Late Imperial Beijing*. Cambridge: Published by the Harvard University Asia Center. Distributed by Harvard University Press, 2005.

Bernstein, Andrew. *Modern Passings: Death Rites, Politics, and Social Change in Imperial Japan*. Honolulu: University of Hawai'i Press, 2006.

Boan huibian [Collection of rejected cases]. Beijing: Falü chubanshe, 2009 (1883).

Bodde, Derk. "Forensic Medicine in Pre-Imperial China." *Journal of the American Oriental Society* 102, no. 1 (1982): 1–15.

Bray, Francesca. "Science, Technique, Technology: Passages between Matter and Knowledge in Imperial Chinese Agriculture." *The British Journal for the History of Science* 41, no. 3 (2008): 319–44.

Broman, Thomas. "The Semblance of Transparency: Expertise as a Social Good and an Ideology in Enlightened Societies." *Osiris* 27, no. 1, Clio Meets Science: The Challenges of History (2012): 188–208.

Brook, Timothy, Jérôme Bourgon, and Gregory Blue. *Death by a Thousand Cuts.* Cambridge: Harvard University Press, 2008.

Bullock, Mary Brown. *An American Transplant: The Rockefeller Foundation and Peking Union Medical College.* Berkeley: University of California Press, 1980.

Buoye, Thomas. "Suddenly Murderous Intent Arose: Bureaucratization and Benevolence in Eighteenth-century Qing Homicide Reports." *Late Imperial China* 16, no. 2 (1995): 62–97.

Burney, Ian A. *Bodies of Evidence: Medicine and the Politics of the English Inquest, 1830–1926.* Baltimore: The Johns Hopkins University Press, 2000.

Poison, Detection, and the Victorian Imagination. Manchester: Manchester University Press, 2006.

Burney, Ian A. and Neil Pemberton. "Bruised Witness: Bernard Spilsbury and the Performance of Early Twentieth-Century English Forensic Pathology." *Medical History* 55, no. 1 (2011): 41–60.

Murder and the Making of English CSI. Baltimore: The Johns Hopkins University Press, 2016.

Buzhu xiyuan lu jizheng [Records on the washing away of wrongs with collected evidence, with supplements and annotation]. Beizhi wenchanghui, 1904.

Cai Hongyuan. *Minguo fagui jicheng* [Collection of laws and regulations in Republican China]. Hefei: Huangshan shushe, 1999.

Campbell, Cameron Dougall. "Chinese Mortality Transitions: The Case of Beijing, 1700–1990." PhD diss., University of Pennsylvania, 1995.

Carroll, Peter J. "Fate-Bound Mandarin Ducks: Newspaper Coverage of the 'Fashion' for Suicide in 1931 Suzhou." *Twentieth-Century China* 31, no. 2 (2006): 70–96.

Chang Che-chia. "'Zhongguo chuantong fayixue' de zhishi xingge yu caozuo mailuo" [Knowledge and Practice in "Traditional Chinese Forensic Medicine"]. *Zhongyang yanjiuyuan jindaishi yanjiusuo jikan* 44 (2004): 1–30.

Chang Renchun. *Lao Beijing de fengsu* [Customs of old Beijing]. Beijing: Beijing yanshan chubanshe, 1996.

Chen Chong-Fang. "'Xiyuan lu' zai Qingdai de liuchuan, yuedu yu yingyong" [The Circulation, Reading, and Using of Xiyuan-lu in Qing Dynasty]. *Fazhi shi yanjiu* 25 (2014): 37–94.

Chen, Janet Y. *Guilty of Indigence: The Urban Poor in China, 1900–1953.* Princeton: Princeton University Press, 2012.

Chen Kangyi. "Daonian wo jing'ai de laoshi – Lin Ji jiaoshou" [Mourning my revered and beloved teacher – Professor Lin Ji]. *Zhongguo fayixue zazhi* 6, no. 4 (1991): 233–7.

"Zhongguo fayixue shi" [A history of legal medicine in China]. *Yishi zazhi* 4, no.1 (1952): 1–8.

Chen, Li. *Chinese Law in Imperial Eyes: Sovereignty, Justice, and Transcultural Politics.* New York: Columbia University Press, 2016.

"Legal Specialists and Judicial Administration in Late Imperial China, 1651–1911." *Late Imperial China* 33, no. 1 (2012): 1–54.

Chen Yuan. *Chen Yuan zaonian wenji* [A collection of the early writings of Chen Yuan]. Taipei: Zhongyang yanjiuyuan Zhongguo wenzhe yanjiusuo, 1992.

Chongkan buzhu Xiyuan lu jizheng [Records on the washing away of wrongs with collected evidence, with supplements and annotation, reprinted]. Yuedong shengshu, 1865.

Ch'ü, T'ung-tsu. *Local Government in China under the Ch'ing*. Cambridge: Council on East Asian Studies, Harvard University; distributed by Harvard University Press, 1988.

Chu Minyi and Song Xingcun. "Chi zhenggu zhi huangmiu" [Denouncing the absurdity of steaming bones]. *Yiyao Pinglun*, no. 36 (1930): 1–4.

Cole, James H. "Social Discrimination in Traditional China: The To-Min of Shaohsing." *Journal of the Economic and Social History of the Orient* 25, no. 1 (1982): 100–11.

Cole, Simon A. *Suspect Identities: A History of Fingerprinting and Criminal Identification*. Cambridge: Harvard University Press, 2002.

Crawford, Catherine. "Legalizing Medicine: Early Modern Legal Systems and the Growth of Medico-Legal Knowledge." In *Legal Medicine in History*, edited by Michael Clark and Catherine Crawford. Cambridge: Cambridge University Press, 1994.

Croizier, Ralph C. *Traditional Medicine in Modern China: Science, Nationalism, and the Tensions of Cultural Change*. Cambridge: Harvard University Press, 1968.

Crossland, Zoë. "Of Clues and Signs: The Dead Body and Its Evidential Traces." *American Anthropologist* 111, no. 1 (2009): 69–80.

Cunningham, Andrew and Perry Williams ed. *The Laboratory Revolution in Medicine*. Cambridge: Cambridge University Press, 1992.

Daqing fagui daquan [Compendium of laws and regulations of the Great Qing]. Zhengxue she, n.d. Includes *Daqing fagui daquan xubian*.

de Bary, Wm. Theodore and Irene Bloom. *Sources of Chinese Tradition: From Earliest Times to 1600*, 2nd edition. New York: Columbia University Press, 1999.

De Renzi, Silvia. "Medical Expertise, Bodies, and the Law in Early Modern Courts." *Isis* 98, no. 2 (2007): 315–22.

"Witnesses of the Body: Medico-Legal Cases in Seventeenth-century Rome." *Studies in History and Philosophy of Science* 33 (2002): 219–42.

Despeux, Catherine. "The Body Revealed: The Contribution of Forensic Medicine to Knowledge and Representation of the Skeleton in China." In *Graphics and Text in the Production of Technical Knowledge in China*, edited by Francesca Bray, Vera Dorofeeva-Lichtmann, and Georges Métailié. Leiden: Brill, 2007.

Dierig, Sven, Jens Lachmund, and J. Andrew Mendelsohn. "Introduction: Toward an Urban History of Science." *Osiris*, 2nd Series, Vol. 18, Science and the City (2003): 1–19.

Dikötter, Frank, Lars Laamann, and Zhou Xun. *Narcotic Culture: A History of Drugs in China*. Chicago: The University of Chicago Press, 2004.

Ding Fubao and Xu Yunxuan. *Jinshi fayixue* [Modern legal medicine]. Shanghai: Wenming shuju, 1911.

Dong, Madeleine Yue. "Communities and Communication: A Study of the Case of Yang Naiwu, 1873–1877." *Late Imperial China* 16, no. 1 (1995): 79–119.

Republican Beijing: The City and Its Histories. Berkeley: University of California Press, 2003.

Dray-Novey, Alison. "Spatial Order and Police in Imperial Beijing." *The Journal of Asian Studies* 52, no. 4 (1993): 885–922.

"The Twilight of the Beijing Gendarmerie, 1900–1924." *Modern China* 33, no. 3 (2007): 349–76.

Duncan, Robert Moore. *Peiping Municipality and the Diplomatic Quarter.* Tientsin: Peiyang Press, Ltd., 1933.

Edmond, Gary. "The Law-Set: The Legal-Scientific Production of Medical Propriety." *Science, Technology, & Human Values* 26, no. 2 (2001): 191–226.

Elman, Benjamin A. *On Their Own Terms: Science in China, 1550–1900.* Cambridge: Harvard University Press, 2005.

Eyferth, Jacob. *Eating Rice from Bamboo Roots: The Social History of a Community of Handicraft Papermakers in Rural Sichuan, 1920–2000.* Cambridge: Harvard University Asia Center. Distributed by Harvard University Press, 2009.

Fahmy, Khaled. "The Anatomy of Justice: Forensic Medicine and Criminal Law in Nineteenth-Century Egypt." *Islamic Law and Society* 6, no. 2, The Legal History of Ottoman Egypt (1999): 224–71.

Fan, Fa-ti. *British Naturalists in Qing China: Science, Empire, and Cultural Encounter.* Cambridge: Harvard University Press, 2004.

"Science, State, and Citizens: Notes from Another Shore." *Osiris* 27, no.1, Clio Meets Science: The Challenges of History (2012): 227–49.

Farquhar, Judith. *Knowing Practice: The Clinical Encounter of Chinese Medicine.* Boulder: Westview Press, 1994.

"Fayi yanjiusuo chengli zaiji" [The Research Institute of Legal Medicine will be established shortly]. *Zhonghua faxue zazhi* no. 8 (1931): 101–2.

"Fayi yanjiusuo juxing erzhou jinian Lin suozhang baogao jingguo" [The Research Institute of Legal Medicine observes its second anniversary, Director Lin reports on its progress]. *Zhonghua yixue zazhi* 20, no. 8 (1934): 1101–2.

Fei Jingzhong. *Duan Qirui.* Jindai Zhongguo shiliao congkan di 90 ji. Taipei: Wenhai chubanshe, n.d.

Freidson, Eliot. *Professional Powers: A Study of the Institutionalization of Formal Knowledge.* Chicago: The University of Chicago Press, 1986.

Frey, Gottfried. *Public Health Services in Germany.* n.p.: League of Nations, Health Organisation, preface 1924.

Gaensslen, R.E. *Sourcebook in Forensic Serology, Immunology, and Biochemistry.* Washington: U.S. Department of Justice, National Institute of Justice, 1983.

Gamble, Sidney D. *How Chinese Families Live in Peiping.* New York: Funk & Wagnalls Company, 1933.

Peking: A Social Survey. New York: George H. Doran Company, 1921.

Gluck, Carol. "The End of Elsewhere: Writing Modernity Now." *American Historical Review* 116, no. 3 (2011): 676–87.

Golan, Tal. *Laws of Men and Laws of Nature: The History of Scientific Expert Testimony in England and America.* Cambridge: Harvard University Press, 2004.

Gong Shucang. "Weishenme bu she caipanyi?" [Why not establish forensic medicine experts?]. *Minguo yixue zazhi* 3, no. 5 (1925): 229–30.

Gong Yibing. "Beijing jindai jingcha zhidu zhi quhua yanjiu" [Research on the District Divisions of the Modern Police in Beijing]. *Beijing shehui kexue* no. 4 (2004): 104–14.

Goodman, Bryna. "Being Public: The Politics of Representation in 1918 Shanghai." *Harvard Journal of Asiatic Studies* 60, no. 1 (2000): 45–88.

"The New Woman Commits Suicide: The Press, Cultural Memory, and the New Republic." *The Journal of Asian Studies* 64, no. 1 (2005): 67–101.

Guoli Beijing daxue yixueyuan zhijiaoyuan xuesheng tongxinlu [Directory of administrative and teaching staff and students of National Beijing University Medical School]. Beijing: National Beijing University Medical School, 1940.

Hallam, Elizabeth. "Articulating bones: an epilogue." *Journal of Material Culture* 15, no. 4 (2010): 465–92.

Han Guanghui. *Beijing lishi renkou dili* [Historical demographic geography of Beijing]. Beijing: Beijing daxue chubanshe, 1996.

Han Jianping. "Huashe tianzu: Qingdai jiangutu zhong de pianzhi guge" ["Drawing Legs for a Snake": The Superfluous Bones in the Qing Dynasty Bone Inspection Diagrams]. *Kexue wenhua pinglun* 8, no. 6 (2011): 58–67.

Han Yanlong and Su Yigong. *Zhongguo jindai jingcha shi* [A history of modern policing in China]. Beijing: Shehui kexue wenxian chubanshe, 2000.

Hansson, Anders. *Chinese Outcasts: Discrimination & Emancipation in Late Imperial China*. Leiden: Brill, 1996.

Harrison, Henrietta. "Narcotics, Nationalism and Class in China: The Transition from Opium to Morphine and Heroin in Early Twentieth-Century Shanxi." *East Asian History* no. 32/33 (2006/2007): 151–76.

Heinrich, Larissa. *The Afterlife of Images: Translating the Pathological Body between China and the West*. Durham: Duke University Press, 2008.

Henriot, Christian. "'Invisible Deaths, Silent Deaths': 'Bodies without Masters' in Republican Shanghai." *Journal of Social History* 43, no. 2 (2009): 407–37.

Hepburn, J.C. *A Japanese and English Dictionary; with an English and Japanese Index*. Shanghai: American Presbyterian Mission Press, 1867.

Hilaire-Perez, Liliane and Catherine Verna. "Dissemination of Technical Knowledge in the Middle Ages and the Early Modern Era: New Approaches and Methodological Issues." *Technology and Culture* 47, no. 3 (2006): 536–65.

Hou Yangfang. "Minguo shiqi quanguo renkou tongji shuzi de laiyuan" [On the sources of national population statistical figures during the Republican period]. *Lishi yanjiu*, no. 4 (2000): 3–16.

Hou Yuwen. "Zhi Sifa bu zhi chengwen" [A petition sent to the Ministry of Justice]. *Minguo yixue zazhi* 4, no. 1 (1926): 2–3.

Hsu, Elisabeth. *The Transmission of Chinese Medicine*. Cambridge: Cambridge University Press, 1999.

Hu Qifei. "Xianxing yanduanshu pinglun ji xiugai zhi chuyi" [An assessment of the corpse examination forms currently in effect and a proposal for revision]. *Fayixue jikan* 1, no. 3 (1936): 49–90.

Hu Yifeng. "Bingyin yixue she chutan – chengli beijing, zaoqi huodong yu lishi yiyi" [An initial study of the Bingyin Medical Society – Its background, early activities and historical significance]. *Beijing dang'an shiliao* no. 3 (2005): 184–96.

Huang Qinglan. *Ouhai guanzheng lu* [A record of governance in Ouhai Circuit]. Taipei: Wenhai chubanshe, n.d. (prefaces 1921).

Huang Ruiting. *Fayi qingtian: Lin Ji fayi shengya lu* [A righteous medico-legal expert: A record of the career of the medico-legal expert Lin Ji]. Beijing: Shijie tushu chuban gongsi, 1995.

Zhongguo jinxiandai fayixue fazhan shi [A history of the development of legal medicine in modern China]. Fuzhou: Fujian jiaoyu chubanshe, 1997.

Huang Yuansheng. "Jindai xingshi susong de shengcheng yu zhankai: Daliyuan guanyu xingshi susong chengxu panjue jianshi (1912–1914)" [The emergence and development of modern criminal litigation: Judgments and commentary of the Supreme Court regarding criminal litigation procedures (1912–1914)]. *Qinghua faxue* 8, Special issue: Research on codification (2006): 83–133.

Hudecek, Jiri. *Reviving Ancient Chinese Mathematics: Mathematics, History and Politics in the Work of Wu Wen-Tsun*. Abingdon: Routledge, 2014.

Hume, Edward H. "The Contributions of China to the Science and Art of Medicine." *Science* 59, no. 1529 (1924): 345–50.

Javers, Quinn. "The Logic of Lies: False Accusation and Legal Culture in Late Qing Sichuan." *Late Imperial China* 35, no. 2 (2014): 27–55.

Jentzen, Jeffrey M. "Death and Empire: Legal Medicine in the Colonization of India and Africa." In *Medicine and Colonialism: Historical Perspectives in India and South Africa*, edited by Poonam Bala. London: Pickering & Chatto, 2014.

Death Investigation in America: Coroners, Medical Examiners, and the Pursuit of Medical Certainty. Cambridge: Harvard University Press, 2009.

Jia Jingtao. *Shijie fayixue yu fakexue shi* [The world history of legal medicine and sciences]. Beijing: Kexue chubanshe, 2000.

"Xinhai geming yihou de zhongguo fayixue" [Legal medicine in China after the revolution of 1911]. *Zhonghua yishi zazhi* 16, no. 4 (1986): 205–9.

Zhongguo gudai fayixue shi [A history of legal medicine in ancient China]. Beijing: Qunzhong chubanshe, 1984.

"Jiancha zhidu cunfei wenti" [On the question of keeping or discarding the procuratorial system]. *Sifa chucaiguan jikan*, no. 1 (1927): 97–102.

Jiang Shaoyuan. "Lianhe qilai yonghu poushi" [Unite to support the autopsying of cadavers]. *Yixue zhoukan ji* 4 (1931): 268–70.

Jiang Zhenxun. "Diaocha sifa shengzhong ying zhuyi fayi zhi wujian" [My opinion that legal medicine should be paid attention to amidst the investigation of the judiciary]. *Minguo yixue zazhi* 4, no. 2 (1926): 43–6.

Jiangsu shengli sibian [Provincial regulations of Jiangsu, fourth collection]. Jiangsu shuju, 1890.

"Jiangsu Wuxi Liuan zhi huizhi" [Collected records from the case of Liu Lianbin from Wuxi, Jiangsu]. *Minguo yixue zazhi* 2, no.1 (1924): 46–54.

Jing Junjian. *Qingdai shehui de jianmin dengji* [The "mean" social stratum in Qing society]. Hangzhou: Zhejiang renmin chubanshe, 1993.

Johnson, David T. *The Japanese Way of Justice: Prosecuting Crime in Japan.* Oxford: Oxford University Press, 2002.

Johnson, Tina Phillips. *Childbirth in Republican China: Delivering Modernity.* Lanham: Lexington Books, 2011.

Kaviraj, Sudipta. "Modernity and Politics in India." *Daedalus* 129, no. 1, Multiple Modernities (2000): 137–62.

Kinkley, Jeffrey. *Chinese Justice, the Fiction: Law and Literature in Modern China.* Stanford: Stanford University Press, 2000.

Ko, Dorothy. *Cinderella's Sisters: A Revisionist History of Footbinding.* Berkeley: University of California Press, 2005.

Köll, Elisabeth. "The Making of the Civil Engineer in China: Knowledge Transfer, Institution Building, and the Rise of a Profession." In *Knowledge Acts in Modern China: Ideas, Institutions, and Identities,* edited by Robert Culp, Eddy U, and Wen-hsin Yeh. Berkeley: Institute for East Asian Studies Publications, forthcoming.

Kremer, Richard L. "Building Institutes for Physiology in Prussia, 1836–1846: Contexts, Interests and Rhetoric." In *The Laboratory Revolution in Medicine,* edited by Andrew Cunningham and Perry Williams. Cambridge: Cambridge University Press, 1992.

Kselman, Thomas A. *Death and the Afterlife in Modern France.* Princeton: Princeton University Press, 1993.

Kuo, Margaret. *Intolerable Cruelty: Marriage, Law, and Society in Early Twentieth-Century China.* Lanham: Rowman & Littlefield Publishers, Inc., 2012.

Kwok, D.W.Y. *Scientism in Chinese Thought, 1900–1950.* New Haven: Yale University Press, 1965.

Lam, Tong. *A Passion for Facts: Social Surveys and the Construction of the Chinese Nation-state, 1900–1949.* Berkeley: University of California Press, 2011.

"Policing the Imperial Nation: Sovereignty, International Law, and the Civilizing Mission in Late Qing China." *Comparative Studies in Society and History* 52, no. 4 (2010): 881–908.

Laqueur, Thomas. "Bodies, Death, and Pauper Funerals." *Representations* no. 1 (1983): 109–31.

Latour, Bruno. *We Have Never Been Modern.* Translated by Catherine Porter. Cambridge: Harvard University Press, 2002.

Lean, Eugenia. "Proofreading Science: Editing and Experimentation in Manuals by a 1930s Industrialist." In *Science and Technology in Modern China, 1880s–1940s,* edited by Jing Tsu and Benjamin A. Elman. Leiden: Brill, 2014.

Public Passions: The Trial of Shi Jianqiao and the Rise of Popular Sympathy in Republican China. Berkeley: University of California Press, 2007.

Lee, Leo Ou-fan and Andrew J. Nathan. "The Beginnings of Mass Culture: Journalism and Fiction in the Late Ch'ing and Beyond." In *Popular Culture in Late Imperial China,* edited by David Johnson, Andrew J. Nathan, and Evelyn S. Rawski. Berkeley: University of California Press, 1985.

Lee, Robert H.G. *The Manchurian Frontier in Ch'ing History.* Cambridge: Harvard University Press, 1970.

Lee, Sophia. "Education in Wartime Beijing: 1937–1945." PhD diss., University of Michigan, 1996.

Lei, Sean Hsiang-lin. *Neither Donkey nor Horse: Medicine in the Struggle over China's Modernity*. Chicago: The University of Chicago Press, 2014.

Lenoir, Timothy. *Instituting Science: The Cultural Production of Scientific Disciplines*. Stanford: Stanford University Press, 1997.

Li Jiarui. *Beiping fengsu leizheng* [A categorized collection of Beiping customs]. Shanghai: Shangwu yinshuguan, 1937.

Li Shizhen. *Bencao gangmu: Xin jiaozhu ben* [Systematic Materia Medica, A new critically-annotated edition]. Edited by Liu Hengru and Liu Shanyong. Beijing: Huaxia chubanshe, 2002.

Li Tao. "Jiyi xiuzheng zhi jiepou shiti guize" [On the necessity of revisions to the *Regulations on the dissection of corpses*]. *Zhonghua yixue zazhi* 16, no. 6 (1930): 529–34.

Li Xiuyun. "*Dagongbao*" *zhuankan yanjiu* [Research on specialized supplements in "Dagongbao"]. Beijing: Xinhua chubanshe, 2007.

Li Yuanxin (William Yinson Lee). *Huanqiu Zhongguo mingren zhuanlüe: Shanghai gongshang gejie zhi bu* [World Chinese Biographies: Shanghai Commercial and Professional Edition]. Shanghai: Globe Publishing Company, 1944.

Liang Qizi. *Shishan yu jiaohua: Mingqing de cishan zuzhi* [Doing good and moral improvement: charitable organizations during the Ming and Qing]. Taipei: Lianjing chuban shiye gongsi, 1997.

Lin Ji. *Fayixue gelun* [A detailed discussion of legal medicine]. Nanjing: Sifa xingzheng bu, 1930.

"Fayixue shilüe" [A brief history of legal medicine]. *Beiping yikan* 4, no. 8 (1936): 22–30.

"Fayixue sizhong xiao shiyan" [Four kinds of small experiments in legal medicine]. *Guoli Beiping daxue yixue niankan* 1, no. 1 (1932): 297–315.

"Niyi chuangli Zhongyang daxue yixueyuan fayi xueke jiaoshi yijianshu" [An opinion regarding the proposed establishment of a legal medicine institute in the medical school of Central University]. *Zhonghua yixue zazhi* 14, no. 6 (1928): 205–16.

"Renlichefu xinzang ji maibo zhi biantai" [Abnormalities in the heart and pulse of rickshaw pullers]. *Zhonghua yixue zazhi* 14, no. 4 (1928): 252–70.

"Shiyong fayixue" [Practical legal medicine]. *Fayi yuekan*, no. 7 (1934): 1–42.

"Sifa gailiang yu fayixue zhi guanxi" [On the relationship between judicial reform and legal medicine]. *Chenbao liu zhou jinian zengkan*, 5th edition (30 January 1925): 48–53.

"Sifa xingzheng bu fayi yanjiusuo chengli yizhounian gongzuo baogao." [Report on the work carried out by the Research Institute of Legal Medicine of the Ministry of Judicial Administration in the year since its founding], *Fayi yuekan*, no. 1 (1934): 1–20.

"Zuijin fayixuejie jiandingfa zhi jinbu" [Recent progress in medico-legal appraisal methods]. *Zhonghua yixue zazhi* 12, no. 3 (1926): 220–37.

Lin Zexu. *Lin Zexu ji: gongdu* [Collected writings of Lin Zexu: Administrative documents]. Beijing: Zhonghua shuju, 1963.

Lin Zhen'gang. "Liunian jian bingli jiepou zhi tongji di guancha" [A statistical survey from six years of pathological dissections]. *Guoli Beijing yixue zhuanmen xuexiao shizhou jinian lunwenji*. Beijing: National Medical College, 1922.

Linden, Allen B. "Politics and Education in Nationalist China: The Case of the University Council, 1927–1928." *The Journal of Asian Studies* 27, no. 4 (1968): 763–76.

Liu Guangding. *Aiguo zhengyi yi lüshi: Liu Chongyou xiansheng* [A patriotic and justice-seeking lawyer: Mr. Liu Chongyou]. Taibei: Xiuwei zixun keji gufen youxian gongsi, 2012.

Liu Guoming. *Zhongguo Guomindang bainian renwu quanshu* [Biographical compendium of personages from one hundred years of the Nationalist Party]. Beijing: Tuanjie chubanshe, 2005.

Liu, Huwy-min Lucia. "Dying Socialist in Capitalist Shanghai: Ritual, Governance, and Subject Formation in Urban China's Modern Funeral Industry." PhD diss., Boston University, 2015.

Lu Runzhi. "Jiepou shiti guize zhi piping" [Criticisms of the *Regulations on the dissection of corpses*]. *Yiyao pinglun* 5, no. 1 (1933): 11–4.

Luesink, David. "Dissecting Modernity: Anatomy and Power in the Language of Science in China." PhD diss., University of British Columbia, 2012.

"State Power, Governmentality, and the (Mis)remembrance of Chinese Medicine." In *Historical Epistemology and the Making of Modern Chinese Medicine*, edited by Howard Chiang. Manchester: Manchester University Press, 2015.

Lüliguan jiaozheng Xiyuan lu [Records on the washing away of wrongs, edited by the Codification Office]. Undated Qianlong edition. *Xuxiu siku quanshu*. Shanghai: Shanghai guji chubanshe, 1995, v. 972.

Luo Zhufeng. *Hanyu dacidian*. Shanghai: Shanghai cishu chubanshe, 2008.

Lynch, Michael, Simon A. Cole, Ruth McNally, and Kathleen Jordan. *Truth Machine: The Contentious History of DNA Fingerprinting*. Chicago: The University of Chicago Press, 2008.

Macauley, Melissa. *Social Power and Legal Culture: Litigation Masters in Late Imperial China*. Stanford: Stanford University Press, 1998.

McKnight, Brian E. *The Washing Away of Wrongs: Forensic Medicine in Thirteenth-Century China*. Ann Arbor: Center for Chinese Studies, The University of Michigan, 1981.

Medical Reports for the half year ended 30th September 1872; Forwarded by the surgeons to the customs at the treaty ports in China; Being no. 4 of the series, and forming the sixth part of the Customs Gazette for July–September 1872. Published by order of the Inspector General of Customs. Shanghai: Printed at the Customs Press, 1873.

Meng Liye. *Zhongguo gong'an xiaoshuo yishu fazhan shi* (A history of the development of the art of court-case fiction in China). Beijing: Jingguan jiaoyu chubanshe, 1996.

Meng Yue. *Shanghai and the Edges of Empires*. Minneapolis: University of Minnesota Press, 2006.

Meyer-Fong, Tobie. *What Remains: Coming to Terms with Civil War in 19th Century China*. Stanford: Stanford University Press, 2013.

Ming Zhongqi. "Woguo fayi qiantu de zhanwang" [A view of the future prospects for legal medicine in our country]. *Dongfang zazhi* 33, no. 7 (1936): 181–6.

Mitchell, Allan. "The Paris Morgue as a Social Institution in the Nineteenth Century." *Francia* 4 (1976): 581–96.

Mnookin, Jennifer L. "Scripting Expertise: The History of Handwriting Identification Evidence and the Judicial Construction of Reliability." *Virginia Law Review* 87, No. 8, Symposium: New Perspectives on Evidence (2001): 1723–845.

Mohr, James C. *Doctors and the Law: Medical Jurisprudence in Nineteenth-Century America*. Baltimore: The Johns Hopkins University Press, 1993.

Morse, Hosea Ballou. *The Gilds of China*. London: Longmans, Green and Co., 1909.

Nam, Yun Sik, Sung-Ok Won, and Kang-Bong Lee. "Modern Scientific Evidence Pertaining to Criminal Investigations in the Chosun Dynasty Era (1392–1897 A.C.E.) in Korea." *Journal of Forensic Sciences* 59, no. 4 (2014): 974–7.

Nanba Mokusaburō. *Fanzui soucha fa* [Methods of criminal investigation]. Translated by Xu Suzhong. Shanghai: Shanghai faxue bianyi she, 1933.

Nappi, Carla. *The Monkey and the Inkpot: Natural History and Its Transformations in Early Modern China*. Cambridge: Harvard University Press, 2009.

Naquin, Susan. "Funerals in North China: Uniformity and Variation." In *Death Ritual in Late Imperial and Modern China*, edited by James L. Watson and Evelyn S. Rawski. Berkeley: University of California Press, 1988.

 Peking: Temples and City Life, 1400–1900. Berkeley: University of California Press, 2000.

Nedostup, Rebecca. *Superstitious Regimes: Religion and the Politics of Chinese Modernity*. Cambridge: Published by the Harvard University Asia Center. Distributed by Harvard University Press, 2009.

Neighbors, Jennifer M. "The Long Arm of Qing Law? Qing Dynasty Homicide Rulings in Republican Courts." *Modern China* 35, no. 1 (2009): 3–37.

Ng, Michael H.K. *Legal Transplantation in Early Twentieth-Century China: Practicing Law in Republican Beijing (1910s–1930s)*. London: Routledge, 2014.

Ocko, Jonathan K. "'I'll Take It All the Way to Beijing: Capital Appeals in the Qing." *The Journal of Asian Studies* 47, no. 2 (1988): 291–315.

Pearson, Quentin A. "Bodies Politic: Civil Law & Forensic Medicine in Colonial Era Bangkok." PhD diss., Cornell University, 2014.

Pott, F.L. Hawks. *A Short History of Shanghai*. Shanghai: Kelly & Walsh, Ltd., 1928.

Prior, Lindsay. *The Social Organization of Death: Medical Discourse and Social Practices in Belfast*. New York: St. Martin's Press, 1989.

Qinding Da Qing huidian shili. Compiled by Kungang et al. (1899). Reprinted in *Xuxiu siku quanshu*. Shanghai: Shanghai guji chubanshe, 1995.

Quanguo sifa huiyi huibian [Collected materials of the National Judicial Conference]. n.p.:1935.

Quanguo zhengxie wenshi ziliao weiyuanhui. *Wenshi ziliao cungao xuanbian: shehui* [Selection of preserved manuscripts of literary and historical materials. Volume 25: Society]. Beijing: Zhongguo wenshi chubanshe, 2002.

Rawski, Evelyn Sakakida. *Education and Popular Literacy in Ch'ing China.* Ann Arbor: The University of Michigan Press, 1979.

Reed, Bradly. *Talons and Teeth: County Clerks and Runners in the Qing Dynasty.* Stanford: Stanford University Press, 2000.

Reed, Christopher. *Gutenberg in Shanghai: Chinese Print Capitalism, 1876–1937.* University of Hawai'i Press, 2004.

Reeves, Caroline. "Grave Concerns: Bodies, Burial, and Identity in Early Republican China." In *Cities in Motion: Interior, Coast, and Diaspora in Transnational China,* edited by David Strand, Sherman Cochran, and Wen-hsin Yeh. Berkeley: Institute of East Asian Studies, University of California, 2007.

Report of the Commission on Extraterritoriality in China. Washington: Government Printing Office, 1926.

Reynolds, Douglas R. *China, 1898–1912: The Xinzheng Revolution and Japan.* Cambridge: Council on East Asian Studies, Harvard University. Harvard University Press, 1993.

Richardson, Ruth. *Death, Dissection, and the Destitute.* Chicago: The University of Chicago Press, 2000 (first published 1987).

The Rockefeller Foundation, Division of Medical Education. *Methods and Problems of Medical Education (Ninth Series).* New York: The Rockefeller Foundation, 1928.

Rogaski, Ruth. *Hygienic Modernity: Meanings of Health and Disease in Treaty-Port China.* Berkeley: University of California Press, 2004.

Röhl, Wilhelm. *History of Law in Japan since 1868.* Leiden: Brill, 2005.

Saeki Yūichi, Tanaka Issei, Hamashita Takeshi, and Ueda Shin, eds. *Niida Noboru hakushi shū Pekin kōshō girudo shiryōshū* [A collection of materials on industrial and commercial guilds of Beijing, compiled by Dr. Niida Noboru]. Tokyo: Tōkyō daigaku Tōyō bunka kenkyūjo Tōyōgaku bunken sentā, 1975–1983.

Sanyiba yundong ziliao [Materials on the March Eighteenth Movement]. Beijing: Renmin chubanshe, 1984.

Sappol, Michael. *A Traffic of Dead Bodies: Anatomy and Embodied Social Identity in Nineteenth-Century America.* Princeton: Princeton University Press, 2002.

Schäfer, Dagmar. *The Crafting of the 10,000 Things: Knowledge and Technology in Seventeenth-Century China.* Chicago: The University of Chicago Press, 2011.

Schayegh, Cyrus. *Who Is Knowledgeable Is Strong: Science, Class, and the Formation of Modern Iranian Society, 1900–1950.* Berkeley: University of California Press, 2009.

Scheffer, Thomas. "Knowing How to Sleepwalk: Placing Expert Evidence in the Midst of an English Jury Trial." *Science, Technology, & Human Values* 35, no. 5 (2010): 620–44.

Scheid, Volker. *Chinese Medicine in Contemporary China: Plurality and Synthesis.* Durham: Duke University Press, 2002.

Schmalzer, Sigrid. *The People's Peking Man: Popular Science and Human Identity in Twentieth-Century China.* Chicago: The University of Chicago Press, 2008.

Schwartz, Vanessa R. *Spectacular Realities: Early Mass Culture in Fin-de-siècle Paris.* Berkeley: University of California Press, 1998.

Scott, James C. *Seeing Like a State: How Certain Schemes to Improve the Human Condition Have Failed*. New Haven: Yale University Press, 1998.

Shapiro, Hugh. "The Puzzle of Spermatorrhea in Republican China." *positions: east asia cultures critique* 6, no. 3 (1998): 551–96.

"The View from a Chinese Asylum: Defining Madness in 1930s Peking." PhD diss., Harvard University, 1995.

Shen, Grace Yen. *Unearthing the Nation: Modern Geology and Nationalism in Republican China*. Chicago: The University of Chicago Press, 2014.

Shen Zhiqi. *Daqing lü jizhu* [The Great Qing Statutes with compiled commentary]. 1746 edition (1715). *Xuxiu siku quanshu*. Shanghai: Shanghai guji chubanshe, 1995, vol. 863.

Shi gong'an [Cases of Judge Shi]. Beijing: Baowentang shudian, 1982.

Shi, Mingzheng. "Beijing transforms: Urban infrastructure, public works, and social change in the Chinese capital, 1900–1928." PhD diss., Columbia University, 1993.

Sifa bu zongwuting diwuke. *Di yi ci xingshi tongji nianbao* [First annual report of criminal statistics]. N.p.: Gonghe yinshuaju, 1917.

Sifa bu zongwuting diwuke. *Di er ci xingshi tongji nianbao* [Second annual report of criminal statistics]. N.p.: Gonghe yinshuaju, 1918.

Sifa bu zongwuting diwuke. *Di san ci xingshi tongji nianbao* [Third annual report of criminal statistics]. N.p.: Gonghe yinshuaju, 1919.

Sifa bu zongwuting diwuke. *Di si ci xingshi tongji nianbao* [Fourth annual report of criminal statistics]. N.p.: Gonghe yinshuaju, 1921.

Sifa ligui bubian [Supplementary collection of judicial regulations]. Beijing: Sifa gongbao faxingsuo, 1919.

Sifa xingzheng bu Fayi yanjiusuo choubei jingguo qingxing ji xianzai chuli shiwu ji jianglai jihua gailüe [An account of the course and circumstances of preparations for the Research Institute of Legal Medicine of the Ministry of Judicial Administration and summary of work presently handled and future plans]. Shanghai: Fayi yanjiusuo, 1932.

Simonis, Fabien. "Mad Acts, Mad Speech, and Mad People in Late Imperial Chinese Law and Medicine." PhD diss., Princeton University, 2010.

Sivin, Nathan, ed. *Science and Civilisation in China. Volume 6: Biology and Biological Technology. Part VI: Medicine*. By Joseph Needham with the collaboration of Lu Gwei-djen. Cambridge: Cambridge University Press, 2000.

Snyder-Reinke, Jeff. "Afterlives of the Dead: Uncovering Graves and Mishandling Corpses in Nineteenth Century China." *Frontiers of History in China* 11, no. 1 (2016): 1–21.

Sommer, Matthew H. *Sex, Law, and Society in Late Imperial China*. Stanford: Stanford University Press, 2000.

"Some Problems with Corpses: Standards of Validity in Qing Homicide Cases." Paper prepared for "Standards of Validity in Late Imperial China," Cluster of Excellence: Asia and Europe in a Global Context, Heidelberg University, October 2013.

Song, Daren. "Weida fayixuejia Song Ci zhuanlüe" [A brief biography of the great medico-legal expert Song Ci]. *Yixue shi yu baojian zuzhi* 2 (1957): 116–21.

Song tixing Xiyuan jilu [Collected writings on the washing away of wrongs of Judicial Commissioner Song]. Yuan edition. *Xuxiu siku quanshu*. Shanghai: Shanghai guji chubanshe, 1995, vol. 972.

Stapleton, Kristin. *Civilizing Chengdu: Chinese Urban Reform, 1895–1937*. Cambridge: Published by the Harvard University Asia Center and distributed by Harvard University Press, 2000.

Star, Susan Leigh and James R. Griesemer. "Institutional Ecology, 'Translations' and Boundary Objects: Amateurs and Professionals in Berkeley's Museum of Vertebrate Zoology, 1907–1939." *Social Studies of Science* 19 (1989): 387–420.

Strand, David. *Rickshaw Beijing: City People and Politics in the 1920s*. Berkeley: University of California Press, 1989.

Sun Kuifang. "Zaoqi pouyan zhi zhongyao" [The importance of early autopsy]. *Fayi yuekan*, no. 17 (1935): 1–10.

Sutton, Donald S. "Death Rites and Chinese Culture: Standardization and Variation in Ming and Qing Times." *Modern China* 33, no. 1 (2007): 125–53.

Tang Erhe. "Xue fazheng de ren keyi budong xie yixue ma?" [Can those who study law and politics not also understand a bit about medicine?]. *Xin jiaoyu* 2, no. 3 (1919): 295–303.

Tang Tenghan. "Xiyuan lu shang zhi huaxue wenti" [Chemistry problems in the *Washing Away of Wrongs*]. *Guofeng banyuekan* 3, no. 12 (1933), 21–3.

Tao Shanmin. "Yiyuan shiyanshi zhi xingzhi gongzuo ji guanlifa" [On the nature, work, and methods of administering hospital laboratories]. *Zhonghua yixue zazhi* 19, no. 5 (1933): 705–15.

Taylor, Kim. *Chinese Medicine in Early Communist China, 1945–63: A Medicine of Revolution*. Abingdon: RoutledgeCurzon, 2005.

Thompson, Malcolm. "Foucault, Fields of Governability, and the Population–Family–Economy Nexus in China." *History and Theory* 51, no. 1 (2012): 42–62.

Tiffert, Glenn D. "The Chinese Judge: From Literatus to Cadre (1906–1949)." In *Knowledge Acts in Modern China: Ideas, Institutions, and Identities*, edited by Robert Culp, Eddy U, and Wen-hsin Yeh. Berkeley: Institute for East Asian Studies Publications, forthcoming.

"An Irresistible Inheritance: Republican Judicial Modernization and Its Legacies to the People's Republic of China." *Cross-Currents: East Asian History and Culture Review* 7 (2013): 84–112.

Timmermans, Stefan. *Postmortem: How Medical Examiners Explain Suspicious Deaths*. Chicago: The University of Chicago Press, 2006.

Tokyo teikoku daigaku hōigaku kyōshitsu gojūsannen shi [Fifty-three year history of the Tokyo Imperial University Department of Legal Medicine]. Tokyo: Tokyo teikoku daigaku igakubu hōigaku kyōshitsu, 1943.

Vanatta, Paul R. and Charles S. Petty. "Limitations of the Forensic External Examination in Determining the Cause and Manner of Death." *Human Pathology* 18, no. 2 (1987): 170–74.

Wakeman, Frederic, Jr. *Policing Shanghai 1927–1937*. Berkeley: University of California Press, 1995.

Waldron, Arthur. *From War to Nationalism: China's Turning Point, 1924–1925.* Cambridge: Cambridge University Press, 1995.

Waley-Cohen, Joanna. "Politics and the Supernatural in Mid-Qing Legal Culture." *Modern China* 19, no. 3 (1993): 330–53.

Wan, Margaret B. *Green Peony and the Rise of the Chinese Martial Arts Novel.* Albany: State University of New York Press, 2009.

Wang Mingde. *Dulü peixi* [A bodkin for untangling difficulties when reading the Code]. Beijing: Falü chubanshe, 2001 (1674).

Wang, Y. Yvon. "Whorish Representation: Pornography, Media, and Modernity in Fin-de-siècle Beijing." *Modern China* 40, no. 4 (2014): 351–92.

Wang You and Yang Hongtong. *Shiyong fayixue daquan* [Great compendium of practical legal medicine]. Tokyo: Kanda insatsujo, 1909.

Warner, John Harley. "The Fall and Rise of Professional Mystery: Epistemology, Authority, and the Emergence of Laboratory Medicine in Nineteenth-Century America." In *The Laboratory Revolution in Medicine*, edited by Andrew Cunningham and Perry Williams. Cambridge: Cambridge University Press, 1992.

Watson, James L. "Funeral Specialists in Cantonese Society: Pollution, Performance, and Social Hierarchy." In *Death Ritual in Late Imperial and Modern China*, edited by James L. Watson and Evelyn S. Rawski. Berkeley: University of California Press, 1988.

"The Structure of Chinese Funerary Rites: Elementary Forms, Ritual Sequence, and the Primacy of Performance." In *Death Ritual in Late Imperial and Modern China*, edited by James L. Watson and Evelyn S. Rawski. Berkeley: University of California Press, 1988.

Watson, James L. and Evelyn S. Rawski, ed. *Death Ritual in Late Imperial and Modern China.* Berkeley: University of California Press, 1988.

Watson, Katherine D. *Forensic Medicine in Western Society: A History.* New York: Routledge, 2011.

Weston, Timothy B. "Minding the Newspaper Business: The Theory and Practice of Journalism in 1920s China." *Twentieth-Century China* 31, no. 2 (2006): 4–31.

Whyte, Martin K. "Death in the People's Republic of China." In *Death Ritual in Late Imperial and Modern China*, edited by James L. Watson and Evelyn S. Rawski. Berkeley: University of California Press, 1988.

Will, Pierre-Étienne. "Developing Forensic Knowledge through Cases in the Qing Dynasty." In *Thinking with Cases: Specialist Knowledge in Chinese Cultural History*, edited by Charlotte Furth, Judith T. Zeitlin, and Ping-chen Hsiung. Honolulu: University of Hawai'i Press, 2007.

Wiltenburg, Joy. "True Crime: The Origins of Modern Sensationalism." *American Historical Review* 109, no. 5 (2004): 1377–404.

Woodhead, H.G.W. *The China Year Book 1929–30.* Tientsin: The Tientsin Press, 1930.

The China Year Book 1933. Shanghai: The North-China Daily News & Herald, 1933.

Wu, Yi-Li. "Between the Living and the Dead: Trauma Medicine and Forensic Medicine in the Mid-Qing." *Frontiers of History in China* 10, no. 1 (2015): 38–73.

"Bodily Knowledge and Western Learning in Late Imperial China: The Case of Wang Shixiong (1808–68)." In *Historical Epistemology and the Making of Modern Chinese Medicine*, edited by Howard Chiang. Manchester: Manchester University Press, 2015.

Wu Yuanbing, ed. *Shen Wensu gong zhengshu* [Official writings of Shen Wensu]. Taipei: Wenhai chubanshe, 1967 (1880).

Xia Quanyin. *Zhiwen shiyan lu* [A record of practical demonstrations of finger-printing]. Beijing: Zhongguo yinshuju, 1926.

Zhiwen xueshu [The academic learning of fingerprinting]. In *Zhentan congshu* [Collectanea on detection], edited Xia Quanyin et al. Nanjing: Jinghua yinshuguan, 1935.

Xiao, Tie. *Revolutionary Waves: Imagining Crowds in Modern China, 1900–1950.* Book manuscript.

Xie Rucheng. *Qingmo jiancha zhidu ji qi shijian* [The late Qing procuratorial system and its practice]. Shanghai: Shanghai renmin chubanshe, 2008.

Xie, Xin-zhe. "Procedural Aspects of Forensics Viewed through Bureaucratic Literature in Late Imperial China." Paper prepared for "Global Perspectives on the History of Chinese Legal Medicine," University of Michigan, Ann Arbor, October 2011.

"The Shaping of Autopsy Evidence in Nineteenth-Century China." Paper prepared for "The Social Lives of Dead Bodies in Modern China," Brown University, June 2013.

Xu Lian. *Xiyuan lu xiangyi* [Detailed explanations of the meaning of the washing away of wrongs]. 1854 Preface. 1877 edition of the office of the Provincial Administration Commissioner of Hubei. *Xuxiu siku quanshu.* Shanghai: Shanghai guji chubanshe, 1995, v. 972.

Xiyuan lu xiangyi [Detailed explanations of the meaning of the washing away of wrongs]. 1854 Preface. Hubei guanshu chu, 1890.

Xu, Xiaoqun. *Chinese Professionals and the Republican State: The Rise of Professional Associations in Shanghai, 1912–1937.* Cambridge: Cambridge University Press, 2001.

Trial of Modernity: Judicial Reform in Early Twentieth-Century China, 1901–1937. Stanford: Stanford University Press, 2008.

Xu, Yamin. "Wicked Citizens and the Social Origins of China's Modern Authoritarian State: Civil Strife and Political Control in Republican Beiping, 1928–1937." PhD diss., University of California, Berkeley, 2002.

"Xuan Axiang an jianyan jingguo" [Circumstances of the forensic examination in the case of Xuan Axiang]. *Yiyao pinglun*, no. 45 (1930): 24–9.

Xue Yunsheng. *Duli cunyi chongkanben* [A typeset edition of the Tu-Li Ts'un-I with a biography of the compiler and numbering and titles added to the sub-statutes]. Edited by Huang Tsing-chia. Taipei: Chengwen chubanshe, 1970 (1905).

Yang Fengkun. "'Wuzuo' xiaokao" [A brief inquiry into the "wuzuo"]. *Faxue* 7 (1984): 40–1.

Yang Nianqun. *Zaizao "bingren": Zhongxiyi chongtuxia de kongjian zhengzhi, 1832–1985* [Re-Making "Patients"]. Beijing: Zhongguo renmin daxue chubanshe, 2006.

Yang Shangheng. "Jiying tichang zhi shiti pouyan" [On the urgent necessity of promoting the autopsy of corpses]. *Tongji zazhi*, no. 7 (1922): 24–5.

Yang Yuanji. "Fayixue shilüe bu" [Supplement to *A Brief History of Legal Medicine*]. *Beiping yikan* 4, no. 9 (1936): 11–2.

Yeh, Wen-hsin. *The Alienated Academy: Culture and Politics in Republican China, 1919–1937*. Cambridge: Council on East Asian Studies, Harvard University, 1990.

   Shanghai Splendor: Economic Sentiments and the Making of Modern China, 1843–1949. Berkeley: University of California Press, 2007.

"Yige zhizhao de jianyan" [The inspection of a fingernail]. *Beiping yikan* 4, no. 8 (1936): 55–9.

Yin Haijin. *Qingdai jinshi cidian* [A dictionary of *jinshi* degree-holders of the Qing dynasty]. Beijing: Zhongguo wenshi chubanshe, 2004.

Yip, Ka-che. *Health and National Reconstruction in Nationalist China: The Development of Modern Health Services, 1928–1937*. Ann Arbor: Association for Asian Studies, Inc., 1995.

You Jinsheng. "Qingmo Beijing neiwaicheng jumin siyin fenxi" [An analysis of causes of death of residents of the Inner and Outer Cities in Beijing at the end of the Qing]. *Zhonghua yishi zazhi* 24, no. 1 (1994): 23–4.

"Youguan sanyiba can'an sishang qingkuang de dang'an" [Archival materials on the situation of the dead and wounded of the March Eighteenth massacre]. *Beijing dang'an shiliao*, no. 1 (1986): 16–30.

Youxian. "Wei yixue qingyuan yu xinwenjie" [A petition for the journalism profession on behalf of medicine]. *Yixue zhoukan ji* 4 (1931): 253–5.

   "Xiang Jingbao jizhe jin yi yan" [A word to the reporters of *Jingbao*]. *Yixue zhoukan ji* 4 (1931): 257–60.

   "Zhi Henshui xiansheng yi feng gongkai de xin" [An open letter to Mr. Henshui]. *Yixue zhoukan ji* 4 (1931): 255–6.

Yu Jiang. "Sifa chucaiguan chukao" [An initial examination of the Judicial Personnel Training School]. In *Qinghua faxue* 4, Ershi shiji hanyu wenming faxue yu faxuejia yanjiu zhuanhao [Special issue on jurisprudence and jurists in twentieth-century Chinese civilization], edited by Xu Zhangrun. Beijing: Qinghua daxue chubanshe, 2004.

Yu Yan. *Yu Yunxiu zhongyi yanjiu yu pipan* [Yu Yunxiu's research and criticisms regarding Chinese medicine]. Edited by Zu Shuxian. Hefei: Anhui daxue chubanshe, 2006.

"Zenyang zuo fayishi ji fayi zai Zhongguo zhi chulu" [How to be a medico-legal physician and the prospects for legal medicine practitioners in China]. *Fayi yuekan*, no. 6 (1934): 1–4.

Zhang Ning. "Corps et peine capitale dans la Chine impériale: Les dimensions judiciaires et rituelles sous les Ming." *T'oung Pao* 94 (2008): 246–305.

Zhang Ruilin et al. *Beiping zhentan an* [Cases of Beiping detectives]. Beiping: Wenmei shuzhuang, 1932.

Zhang Tingxiang. *Rumu xuzhi wuzhong* [Five works on the essentials of entering the muyou profession]. 1892 Zhejiang shuju edition. Taipei: Wenhai chubanshe, 1968.

Zhang Zaitong and Xian Rijin. *Minguo yiyao weisheng fagui xuanbian, 1912–1948* [Selected laws and regulations on medicine and hygiene in Republican China, 1912–1948]. Jinan: Shandong daxue chubanshe, 1990.

Zheng Zhongxuan. "Lin Ji jiaoshou he tade 'Xiyuan lu boyi'" [Professor Lin Ji and his "Critical disputations on the *Washing Away of Wrongs*"]. *Fayixue zazhi* 7, no. 4 (1991): 145–8.

Zhonggong zhongyang Makesi Engesi Liening Sidalin zhuzuo bianyiju yanjiushi, *Wusi shiqi qikan jieshao, di yi ji* [An introduction to periodicals of the May Fourth period, volume 1]. Beijing: Renmin chubanshe, 1958.

Zhu Hengbi. "Jiepou shiti zhi shangque" [A discussion on the dissection of corpses]. *Zhonghua yixue zazhi* 8, no. 4 (1922): 198–207.

Zhuang Wenya. *Quanguo wenhua jiguan yilan* [An overview of the country's cultural organs]. Shanghai: Shijie shuju, 1934.

Index

A *Categorized Collection of Beiping Customs*
  (Beiping fengsu leizheng), 54
A New Treatise on Anatomy (Quanti xinlun), 169
Advanced School for Police Officials of the
  Ministry of Interior, 141
Ai Shiguang (Ngai Shih-kuang), 170, 171
anatomy, 4, 11, 51n87, 194
  anatomy laws, 162, 165–6, 171, 178,
    182, 184
  and forensic knowledge of inspection
    clerks, 67, 76, 185, 217, 219
  general importance of, 163, 166, 169,
    170, 185
  schemes for obtaining cadavers, 85, 176,
    178, 180
Andrews, Bridie, 10n34
Anhui, 202
Arnold, David, 18
arsenic, 13, 188, 188n8, 203, 206n63, 211
automobiles, accidents involving, 49, 50,
    52, 110–11, 116, 223
autopsy, 23, 159
  criticisms of, 162, 169
  forensic uses of, 3, 21, 80, 121, 124, 125,
    151, 160, 165, 166–8, 193, 198, 208
  medical uses of, 170, 177, 178
  police supervision over, 178, 179
  redundancy of, 154, 163, 184, 185

Beggars' Home, 46n67
Beijing
  administration under the Qing, 42
  city gates, 26, 27, 29, 31, 32, 34, 37, 38,
    43, 47, 47n71, 48, 71, 110, 136
  Japanese occupation of, 85, 216
  layout of, 27
  population of, 27
  suburban districts, 27, 33, 34, 41n53,
    70, 82
  under the Nationalists, 191

Beijing Bar Association, 10n34, 111,
  119–20
Beijing No. 1 Prison, 77, 180, 210
Beijing procuracy, 1, 31
  and city police, 80, 84
  and homicide investigation, 1, 42, 53, 55,
    58, 70–1, 80, 84, 110, 182
  and skeletal examinations, 87, 96,
    102, 187
  criticisms of, 119, 122
  forensic caseload of, 48, 79, 80, 83, 85
Beijing-Hankou Railway, 25, 51
Beijing-Suiyuan Railway, 25, 35
Beijing-Tianjin Railway, 44
Beiping. *See* Beijing
Beiping Bar Association. *See* Beijing Bar
  Association
Beiping Local Court, 1n2, 2, 173
Beiping Medical Journal (Beiping yikan), 209
Beiping Municipal Hospital, 181, 182
Beiping University, 126
Beiping University Medical School, 199,
  211, *See also* National Medical College
  in Beijing
  Institute of Legal Medicine, 2, 16, 125,
    131, 189, 196, 203, 207
Beiping-Mukden Railway, 1, 25
Bertillonage, 13
Bingyin Medical Society, 174–5
blood evidence, 13, 123, 125, 127, 129n41,
    136, 138, 144, 146, 148, 152–4, 155,
    195, 198, 202, 214, 217
Blue, Gregory, 23, 91
Board of Punishments, 42, 72, 104–5
Book of Songs (Shijing), 73
Bourgon, Jérôme, 23, 91
Boxer Uprising, 35, 57
Bray, Francesca, 108
Brook, Timothy, 23, 91
Buddhism, 30, 91
Bureau of Hygiene, 48n75, 84, 181

251

Printed in the United States
by Baker & Taylor Publisher Services

Printed in the United States
by Baker & Taylor Publisher Services